Clean Energy Production Technologies

Series Editors

Neha Srivastava, Department of Chemical Engineering and Technology, IIT (BHU) Varanasi, Varanasi, Uttar Pradesh, India

P. K. Mishra, Department of Chemical Engineering and Technology, IIT (BHU) Varanasi, Varanasi, Uttar Pradesh, India

The consumption of fossil fuels has been continuously increasing around the globe and simultaneously becoming the primary cause of global warming as well as environmental pollution. Due to limited life span of fossil fuels and limited alternate energy options, energy crises is important concern faced by the world. Amidst these complex environmental and economic scenarios, renewable energy alternates such as biodiesel, hydrogen, wind, solar and bioenergy sources, which can produce energy with zero carbon residue are emerging as excellent clean energy source. For maximizing the efficiency and productivity of clean fuels via green & renewable methods, it's crucial to understand the configuration, sustainability and techno-economic feasibility of these promising energy alternates. The book series presents a comprehensive coverage combining the domains of exploring clean sources of energy and ensuring its production in an economical as well as ecologically feasible fashion. Series involves renowned experts and academicians as volume-editors and authors, from all the regions of the world. Series brings forth latest research, approaches and perspectives on clean energy production from both developed and developing parts of world under one umbrella. It is curated and developed by authoritative institutions and experts to serves global readership on this theme.

Neha Srivastava • Bhawna Verma • P. K. Mishra
Editors

Paddy Straw Waste for Biorefinery Applications

Springer

Editors
Neha Srivastava
Department of Chemical Engineering & Technology
Indian Institute of Technology (BHU)
Varanasi, Uttar Pradesh, India

Bhawna Verma
Department of Chemical Engineering & Technology
Indian Institute of Technology (BHU)
Varanasi, Uttar Pradesh, India

P. K. Mishra
Department of Chemical Engineering & Technology
Indian Institute of Technology (BHU)
Varanasi, Uttar Pradesh, India

ISSN 2662-6861 ISSN 2662-687X (electronic)
Clean Energy Production Technologies
ISBN 978-981-99-8223-3 ISBN 978-981-99-8224-0 (eBook)
https://doi.org/10.1007/978-981-99-8224-0

This Springer imprint is published by the registered company Springer Nature Singapore Pte Ltd.
The registered company address is: 152 Beach Road, #21-01/04 Gateway East, Singapore 189721, Singapore

Paper in this product is recyclable.

Preface

This book entitled, *Paddy Straw Waste for Biorefinery Applications*, provides sustainable value addition of paddy straw and its different roles in biorefinery applications. Additionally, the book explores different large-scale industrial uses of paddy straw for different applications along with biorefinery grounds. Huge production, availability, and advantage of the lignocellulosic waste need mass-scale expansion for extensive industrial utilization as well as for effective environmental management. Biotransformation and value addition of paddy straw also help to resolve the issue of solid waste management around the globe because rice straw is included among the most consumed cereal worldwide and releases huge amounts of waste alongside. Therefore, this book gives the common ground to explore this waste for various important biorefinery roles through its maximum utilization. Based on this fact, this book has been divided into ten targeted chapters based on paddy straw application in biorefinery industries. Chapters 1, 2, and 3 present the basic biorefinery concept, importance, and current scenario with paddy straw utility as promising feedstock. Chapters 4, 5, and 6 explore the sustainable possibility of paddy straw in biofuels production via fabrication of nanomaterials and bioethanol production. On the same pattern, Chaps. 7 and 8 explore the charcoal fabrication and current ethanol production development as well as the challenges in the area. Further, Chaps. 9 and 10 discuss the feasibility and improvement in hydrolytic enzyme production and biochar synthesis using paddy straw as the basic and initiated material. Since the book is exclusively based on the sustainable application of paddy straw in biorefinery industries, it will prove its promising credentials in the area to overcome the existing roadblocks and enrich the biorefinery industries.

Contents

Editors and Contributors

About the Editors

Neha Srivastava has received PhD in Biotechnology from the Department of Molecular and Cellular Engineering, SHIATS, India in the area of bioenergy. She is working as Research Scientist in the Department of Chemical Engineering and Technology, IIT (BHU) Varanasi, India. She has published more than 37 research articles in the in peer-reviewed journals of SCI impact factor and have filed three patents, one technology transfer, and 14 published books and two book series of international renowned publisher. She is working on bioprocess technology and biofuels production (microbial screening and enzymes; production and enhancement, biohydrogen production from waste biomass, bioethanol production).

Bhawna Verma is currently working as Associate Professor in the Department of Chemical Engineering and Technology IIT (BHU), Varanasi, India. She got PhD degree from IT-BHU, Varanasi. She has more than 30 potential publications and her area of work involved algal fuel and biodiesel; carbon materials/nanocomposite materials for enhanced capacitance.

P. K. Mishra is currently Professor and Head of the Department of Chemical Engineering and Technology, Indian Institute of Technology (BHU), Varanasi, India. He has authored/co-authored over 60 technical papers published in reputed national/international journals and supervised more than 20 doctoral students. He has received several awards and honours and has five patents with one technology transfer. He is Fellow of Institution of Engineers India. He has received several awards and honours at national/international levels.

Contributors

Suhail Ahmad Department of Bioengineering, Integral University, Lucknow, Uttar Pradesh, India

Abdulrahman H. Alessa Faculty of Science, Department of Biology, University of Tabuk, Tabuk, Saudi Arabia

Shaukat Ali Institute of Zoology, University of the Punjab, Lahore, Pakistan
Department of Zoology, Government College University, Lahore, Pakistan

Mohommad Asif Department of Bioengineering, Integral University, Lucknow, Uttar Pradesh, India

Bindu Bala Punjab Agricultural University, Ludhiana, India

Ravish Choudhary Division of Seed Science and Technology, Indian Agricultural Research Institute, Delhi, India

Priyanka Dubey Department of Bioengineering, Integral University, Lucknow, Uttar Pradesh, India
Department of Biosciences, Integral University, Lucknow, Uttar Pradesh, India

Alvina Farooqui Department of Bioengineering, Integral University, Lucknow, Uttar Pradesh, India

Marcelo Franco Department of Exact Science and Technology, State University of Santa Cruz, Ilheus, Brazil
Department of Exact Science and Technology, State University of Santa Cruz, Ilheus, Brazil

Misbah Ghazanfar Faculty of Science, Department of Biotechnology, University of Sargodha, Sargodha, Pakistan

Alia Gul Department of Botany, Hazara University, Mansehra, Pakistan

Anuja Gupta Punjab School Education Board, Mohali, Punjab, India

Bushra Iqbal Faculty of Science, Department of Biotechnology, University of Sargodha, Sargodha, Pakistan

Muhammad Irfan Faculty of Science, Department of Biotechnology, University of Sargodha, Sargodha, Pakistan
Department of Biotechnology, University of Sargodha, Sargodha, Pakistan

Irum Department of Biosciences, Integral University, Lucknow, Uttar Pradesh, India

Harinderjeet Kaur Department of Food Science, Rayat Bahra University, Mohali, Punjab, India

Manmeet Kaur Department of Microbiology, Punjab Agricultural University, Ludhiana, Punjab, India

Ravneet Kaur Department of Microbiology, Punjab Agricultural University, Ludhiana, Punjab, India

Sandeep Kaur Department of Entomology, Punjab Agricultural University, Ludhiana, Punjab, India

Muhammad Khan Institute of Zoology, University of the Punjab, Lahore, Pakistan

Nortoji A. Khujamshukurov Tashkent Institute of Chemical Technology, Tashkent, Republic of Uzbekistan

G. S. Kocher Department of Microbiology, Punjab Agricultural University, Ludhiana, Punjab, India

Sachin Kumar Food Corporation of India, Ludhiana, Punjab, India

Parnika Mishra Department of Biosciences, Integral University, Lucknow, Uttar Pradesh, India

Vishal Mishra School of Biochemical Engineering, IIT (BHU), Varanasi, Uttar Pradesh, India

Vishnavi Mishra Department of Biosciences, Integral University, Lucknow, Uttar Pradesh, India

Tanvi Sahni Punjab Agricultural University, Ludhiana, Punjab, India

Hafiz Abdullah Shakir Institute of Zoology, University of the Punjab, Lahore, Pakistan

Deeksha Sharma Institute of Biomedical and Research, Mangalaytan University, Aligarh, Uttar Pradesh, India

Neha Sharma Department of Microbiology, Punjab Agricultural University, Ludhiana, Punjab, India

Shivani Sharma Department of Microbiology, Punjab Agricultural University, Ludhiana, Punjab, India

Tarun Sharma Department of Agronomy, CSK Himachal Pradesh Agricultural University, Palampur, Himachal Pradesh, India

Aneeshpal Singh Department of Food Science and Technology, Punjab Agricultural University, Ludhiana, Punjab, India

Diksha Singla Department of Biochemistry, Punjab Agricultural University, Ludhiana, Punjab, India

Aqsa Tahir Department of Biotechnology, University of Sargodha, Sargodha, Pakistan

Gyanendra Tripathi Department of Bioengineering, Integral University, Lucknow, Uttar Pradesh, India

Diksha Verma Punjab Agricultural University, Ludhiana, Punjab, India

Chapter 1
Biorefineries: An Analogue to Petroleum Refineries

Anuja Gupta, Tanvi Sahni, and Sachin Kumar

Abstract This chapter explores the concept of biorefineries as a sustainable and efficient alternative to traditional petroleum refineries. It delves into the increasing global demand for renewable energy resources and the urgent need to transition from fossil fuels to more sustainable options. The abstract begins by highlighting the key objective of the chapter, which is to present biorefineries as a viable solution for the production of various valuable products from biomass feedstocks. It emphasizes the importance of integrating multiple processes in biorefineries to maximize resource utilization and minimize waste generation. The chapter provides an overview of petroleum refineries as a basis for comparison. It highlights the similarities between the two refinery types, such as the conversion of raw materials into valuable products. Additionally, it discusses the significant advantages of biorefineries, including the utilization of renewable biomass resources, reduced greenhouse gas emissions, and the potential for bioproduct diversification. Furthermore the abstract delves into the different conversion technologies employed in biorefineries, such as biochemical, thermochemical, and hybrid processes. It explores various biomass feedstocks, including agricultural residues, energy crops, and algae, and their respective conversion pathways. Moreover, this chapter emphasizes the importance of biorefinery integration with existing industries and infrastructure, highlighting the potential synergies and economic benefits. It also addresses the challenges associated with biorefinery implementation, including feedstock availability, technological advancements, and market competitiveness. Finally, the abstract concludes by summarizing the chapter's key findings and discussing the future prospects of biorefineries. It highlights the role of policy support and research and development in fostering the growth of biorefinery sector and facilitating a sustainable transition toward a biobased economy. Overall, this chapter serves as the comprehensive

A. Gupta
Punjab School Education Board, Mohali, Punjab, India

T. Sahni
Punjab Agricultural University, Ludhiana, Punjab, India

S. Kumar (✉)
Food Corporation of India, Ludhiana, Punjab, India

N. Srivastava et al. (eds.), *Paddy Straw Waste for Biorefinery Applications*, Clean Energy Production Technologies, https://doi.org/10.1007/978-981-99-8224-0_1

introduction to the concept of biorefineries, showcasing their potential as an analogue to petroleum refineries and as essential component of the renewable energy landscape.

Keywords Biorefineries · Feedstock · Petroleum refineries · Thermochemical and hybrid processes

1.1 Introduction

With the advancement in technology and rise in population growth over the past decades (Alalwan et al. 2019), there is a 64% increased demand for transport and chemical energy with every passing day which results in enhanced usage of biofuels up to ~111 million barrels per day (Mb/d) (Sharma and Singh 2017). Liquid fuels have significant role in the economic development around the globe (Panda et al. 2010). Majorly the global population mainly depends on fossil fuels, which results in the emission of greenhouse gases (Ramachandra et al. 2015). The main energy requirements (80%) are currently fulfilled in the industrial and transport sectors with chemical fuels and petroleum resources (Yusuf et al. 2011). Due to this, carbon dioxide emission over the last 50 years has increased from 200 to 414 ppm (Ballantyne et al. 2012). With this rate, this level can reach up to 500 ppm by 2045, which will cause the melting of the polar ice sheets. This in turn results in the rise of the water levels in the sea to several meters, which gives us a valid reason to shift our focus from the fossil fuels and find some renewable source of fuels for energy production to reduce the carbon footprint on the earth. For example, when we blend 10% bioethanol into gasoline, the emission of greenhouse gases is reduced up to a great extent like reduction in emission of CO_2 by 6–10%, CO by 25–30%, NO_x by 5%, and volatile organic compounds by 7%, respectively (Suarez-Bertoa et al. 2015). Till date, oil is the main source of energy production and transportation. As till date the total daily consumption of oil is about 84 million barrels which is likely to be increased by ~116 million barrels by 2030 (Kjärstad and Johnsson 2009). Studies conducted have shown that lignocellulosic biomass can be used as a good alternative as well as renewable source of fossil fuels and can be efficiently used as transportation fuel (Nanda et al. 2015; Kumar et al. 2023). As biomass availability is also limited, it is necessary to use them efficiently. Thus, biorefineries are thought to play an important part in decarbonizing energy (Stegmann et al. 2020). A noticeably reduced amount of GHG emission can be attained by the use of biomass as biofuels because the carbon dioxide released during the incineration of lignocellulosic plants is balanced by the amount used during plant cultivation (Chen et al. 2021). Thus, the production of bioenergy is proven to be the best alternative till date and contributes ~9% of the entire global energy supply (Gielen et al. 2019). During a research, it was found that there is a significant reduction in GHG emissions in South Africa through the use of advanced biofuels which were prepared from lignocellulosic materials mainly constituting sugarcane bagasse and other biomass resources (Ullah et al. 2015). Several other renewable sources of energy can be used as alternative sources for energy production apart from lignocellulosic biofuels (Ho et al. 2014). Such as heat and electricity are generated from wind, solar, and hydropower energy, the requirement of transportation fuel can be fulfilled from lignocellulosic biofuels (Lange 2007). Biofuels are also helpful in decarbonizing the harmful gases which are the by-products of flight and freight industries (McCollum and Yang 2009). Using biomass for fuel production is a good alternative for the improvement of the environment (Groom et al. 2008). The biomass produced in the form of agricultural, industrial, forestry, aquatic, and municipal solid waste creates environmental pollution by emitting toxic gases as it is decomposing with unstrained methods; thus carbon-based solid wastes can be used as an abundant natural source for biofuels (Inyang et al. 2022). To improve the degrading quality of environment, it is necessary to grow biofuel consumption in automobile industry. It will also prove useful in the growth and advancement of agriculture sector, which in turn will produce more biomass for more biofuel production (Demirbas 2008). In such way, lignocellulosic biomass and biofuel production will slowly dominate the fuel industry in few years.

Over time, different researchers proposed different definitions of a biorefinery based on their demand and usage which kept on changing in due course of time (Borras Jr et al. 2016). "Biorefining is the sustainable processing of biomass into a spectrum of marketable products and energy" (Mohan 2016). "A biorefinery is a facility that integrates biomass conversion processes and equipment to produce fuels, power, and chemicals from biomass" (Béligon et al. 2018). Definitions of biorefineries on the basis of the economic importance of forest industry to the biorefineries: "Full utilization of the incoming biomass and other raw materials for simultaneous and economically optimized production of fibers, chemicals, and energy" (Berntsson et al. 2008). "Maximising the economic value from trees," need "an improved business model and corporate transformation" (Berntsson et al. 2008). An overview of biorefinery is given below (Fig. 1.1).

Biorefineries are "Facilities that can combine biomass conversion processes and equipment to generate fuels, power, and new materials in an economically, socially and environmentally sustainable way" (Berntsson et al. 2008). Biorefinery is an evolutionary concept in the fuel world; thus, it is defined as "A complex facility (or network) that involves integral biomass conversion processes to produce a range of products, mainly biofuels, power, materials, food, and feed as well as chemicals and biochemicals based on biomass" (Cardona-Alzate et al. 2020).

A huge variety of biowastes are used by the biorefineries to produce a wide spectrum of fuels, for example, glycerol is processed to obtain high-added second- and third-generation fuel (Ferreira et al. 2019). Crop produce is used as potential material in biorefineries to prevent land competition (Cherubini 2010). Due to the above reasons, two different types of biorefineries are required by taking into account the final products used during the manufacturing of biofuels. Energy-focused biorefineries prioritize the production of biofuels, power, and heat as their primary outputs. These biorefineries generate various forms of energy, including biofuels such as biogas, syngas, hydrogen, bio-methane, bioethanol, FT fuels, bio-oil, biodiesel, charcoal, etc. Additionally, they also produce electricity and

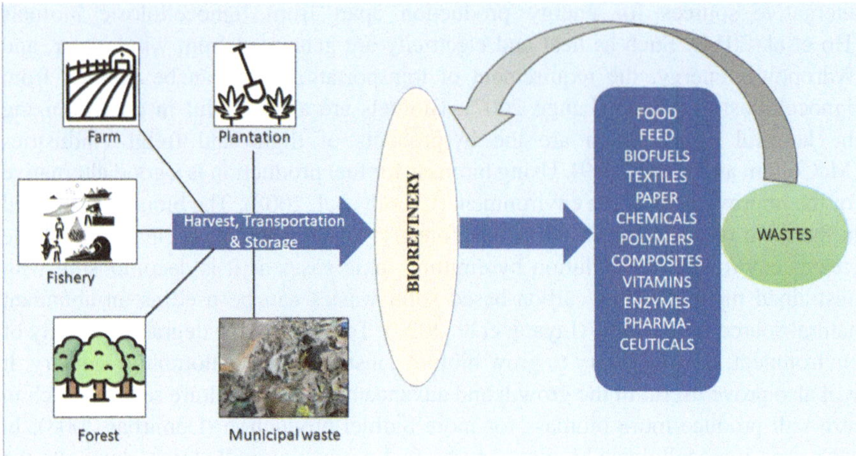

Fig. 1.1 Overview of biorefinery

heat, emphasizing their core purpose of energy generation (Awogbemi et al. 2021), whereas the industries that mainly focus on products are those biorefineries whose main objective is to provide the major products used in energy-driven biorefineries. Such products are:

- Chemicals (fine chemicals, building blocks) (Takkellapati et al. 2018)
- Organic acids (Becker et al. 2015)
- Polymers and resins (Thang and Novalin 2008)
- Biomaterials (Fahd et al. 2012)
- Food and animal feed (Jungmeier 2014)
- Fertilizers (Seghetta et al. 2016)

Bioethanol and biohydrogen have been recognized as beneficial fuels and base chemicals (Gielen et al. 2019; Verma et al. 2023). However, the classification of these products depends on their intended purpose and the demand in the market. In recent times, the concept of biorefineries has emerged, referring to advanced systems that integrate the processing or fractionation of biomass to produce a range of valuable outputs. To achieve this, a comprehensive assessment of the raw materials and the adoption of sustainable design principles are necessary, incorporating the latest advanced technologies and approaches. It is important to note that these considerations align with the three pillars of sustainability, ensuring the viability of the biorefinery in terms of environmental, social, and economic aspects.

1.2 Why Biorefinery Will Replace Oil Refinery?

The industry based on petroleum is unsustainable as the demand for petrochemicals rises day by day due to the finite nature of the fossil fuels from which these chemicals are created, as well as the environmental pollution that results from the overuse of these fossil resources (Bhan et al. 2020). These problems create a need for an alternate sustainable renewable source to meet up the daily energy demands over the globe, which results in an unexpected increase in the development of the biorefinery industry (Kurian et al. 2013). The economy's major material handler in recent years has been the oil industry. Around six billion tons of CO_2 was emitted worldwide as the primary fuel or more than 1000 kg per person (Themelis 2003). In comparison, the global steel industry produces approximately 700 million tons of steel each year, equating to an average of approximately 120 pounds per person. Renewable, photosynthetic biomass should be used as an alternative to oil. Naturally, biomass is preserved in biofuels, and energy is dissipated; oil refineries produce a lot of toxic waste products like phenols, sulfides, and heavy metals, present in crude oil (Wan et al. 2022). During the refining of this crude oil, remaining toxic substances such as cyanide, dioxins, and furans are produced, along with this many other toxic wastes added to the air as well as groundwater, altering their natural composition (Misra and Pandey 2005). The solution to this oil refinery pollution is bioremediation which is one of the methods of biorefinery (Khatiwada et al. 2020). For instance, certain microorganisms are used to break down the pollutants like phenols (Karigar and Rao 2011). The process of refining oil at a petroleum refinery begins with the receipt of crude oil for storage, processing, and shipping of finished products. At the oil refinery, oil corporations utilize catastrophic event risk management (as acceptable risk limits for environmental, public health, and safety). Biomass-based chemicals typically require production procedures with less demanding temperature, pressure, or solvent conditions. Hence the risk is diminished or eliminated via biorefinery (Alfaqiri et al. 2019). In contrast to oil refineries, biomass refineries recycle carbon dioxide. Hence, biorefinery suggests a favorable impact on global warming (Gravitis 1998). The proper functioning of a power plant depends upon how much carbon dioxide is emitted per kWh. Thus the efficiency of electric power plants using wood to generate power is 60% more than the efficiency of the power plants that use coal as the burning material which changes the fossil fuel to plant material ratio to 30:1 (Spliethoff 2010). Several sources of biomass are wasted and disposed of per year in huge amounts, for example, if we talk about the USA, they dispose approximately 350 million tons of agricultural biomass as a waste product every year, and also the biomass production of tropical plantations that is unused is unexpectedly high (Tye et al. 2016). Thus, it is concluded that the world has an enormous amount of residual biomass that can be used as fuel without any harm to the forests and without causing soil erosion. Less than or equal to 40% of the total residual mass from the fields is used for biofuel-ethanol production, and the remaining is again provided back to the fields in the form of organic matter added to the soil which will increase the fertility of the soil, reduce soil erosion, and also remove carbon dioxide in the environment in a noticeable amount (Sinclair and Weiss 2010). Only a 1% rise in the organic matter added to the soil will reduce up to 40 t of carbon dioxide per hectare of land (Winsley 2007). Ethanol transportation reduces up to 46% GHG of the total 25% of greenhouse gas emissions in the USA in place of gasoline (Wang et al. 2011). Thus, to generate full potential from biofuels, the biorefinery industry should be expanded, and engine designs should be improved so that we can take more advantage of the biofuels.

1.3 Refineries vs. Biorefineries

When comparing the manufacturing technologies of biorefineries, it becomes evident that there is a significant and noticeable difference between hydrocarbon-based crude oil and biomass. This disparity in quality can be attributed to the distinct soil varieties from which they are derived, resulting in differing physical and chemical properties of crude oil (Stedile et al. 2015). The properties of biomass, on the other hand, are influenced by factors such as sunlight exposure, maturity duration, and the levels of air and moisture it encounters. These factors contribute to variations in the oxygen content of biomass, which is higher compared to crude oil (Bolan et al. 2021). The potential of biomass as a source for chemical has been extensively studied in various research papers, highlighting its increasing popularity. These biobased products have the potential to generate a significant revenue of more than US$15 billion for the global chemical industry. However, the economic production of transportation biofuels remains a considerable challenge for chemical industries (de Jong et al. 2012). Currently, the fuel industry relies heavily on hydrocarbon products such as LPG and natural gas for transportation, polyester, polyurethane, polymers, glycol, ammonia, synthetic rubber, asphalt, and insecticides (Sudha et al. 2023). Developing biorefineries is aimed at replacing these harmful and finite hydrocarbons with renewable alternatives like biomass-based products. Major differences between refineries and biorefineries are given in Table 1.1.

Figures 1.2 and 1.3 also help in distinguishing between the biorefineries and crude oil based on the different types of products obtained. In crude oil refineries, there is certainty in the composition, technology, and products, but in biorefineries, there is no certainty in the products, technology, location, and composition of the products obtained, although the objective of the biorefinery is the same as that of crude oil refinery (Cardona-Alzate et al. 2020).

In the formation of biorefineries or crude refineries, the location is the main factor for the sustainable operation of the refineries as well as the analysis of the products and their alternatives (DwiPrasetyo et al. 2020). The price of these products is directly influenced by various factors, including the geostrategic policies established by governments and the occurrence of conflicts or wars. Changes in these geopolitical dynamics can disrupt the supply chain, leading to fluctuations in feedstock availability and subsequent price volatility. As a result, the petrochemical industry closely monitors and responds to geopolitical developments as they significantly

Table 1.1 Comparison of refineries and biorefineries

S. No.	Based on	Refinery	Biorefinery
1	Feedstock	Main component of feedstock is typically a hydrocarbon, which contributes to its relative homogeneity	Feedstock is heterogeneous. The bulk components are carbohydrates, lignin, proteins, and oils
		The amount of oxygen content is low	The amount of oxygen content is high
		As processed further the weight of the product is increased	As processed further, the weight of the product is decreased; thus it is important to maintain the amount at the start of the processing to obtain the required quantity of end product
		Sulfur content is high in refinery-based products	Sulfur content is low in biorefinery-based products although it is high in inorganic substances like silica
2	Building block composition	Building blocks of refinery products are hydrocarbons: ethylene, propylene, methane, benzene, toluene, and xylene isomers	The fundamental building blocks involved in various biological processes are glucose, xylose, and fatty acids. These building blocks play crucial roles in biochemical pathways and are essential for the synthesis of diverse compounds. Examples of fatty acids include oleic acid, stearic acid, and sebacic acid, each serving distinct functions within biological systems
3	(Bio)chemical processes	The process primarily involves chemical reactions, particularly the introduction of heteroatoms such as oxygen (O), nitrogen (N), and sulfur (S). This includes techniques like steam cracking and catalytic reforming, which enable a wide range of conversion chemistries to take place. These chemical processes play a significant role in transforming the composition and properties of the substances involved, allowing for the synthesis of various valuable products	Biorefinery operations involve the integration of chemical and biotechnological processes. These processes are employed to remove oxygen from the feedstock and undergo various relative heterogeneous processes, such as dehydration, hydrogenation, and fermentation. Through these transformations, the aim is to obtain building blocks that can be further utilized for the production of desired products. These processes are essential for converting the feedstock into bioenergy, biofuels, chemicals, and high-value compounds
4	Chemical intermediates produced at a commercial scale	Produced in a wide range and huge amount	Produced in fewer amount but its number is increasing with more advancement

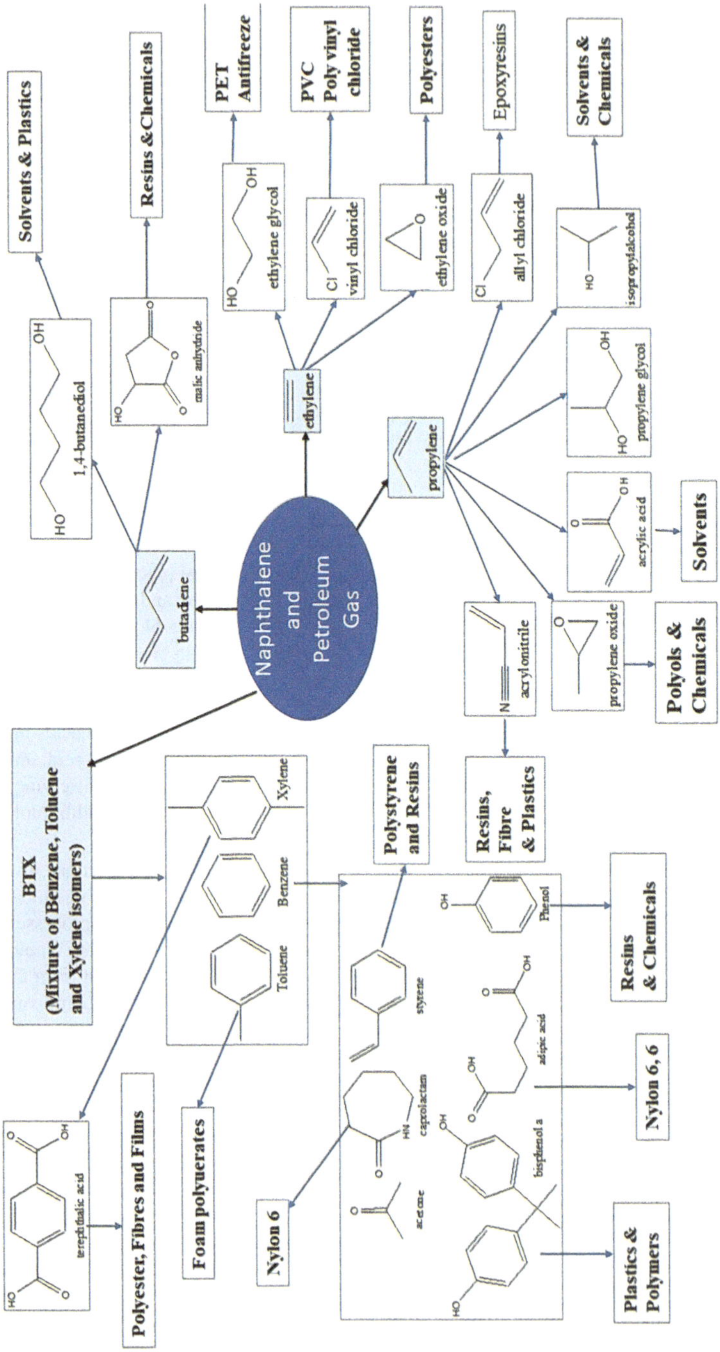

Fig. 1.2 Basic petrochemicals in refinery, and its major applications

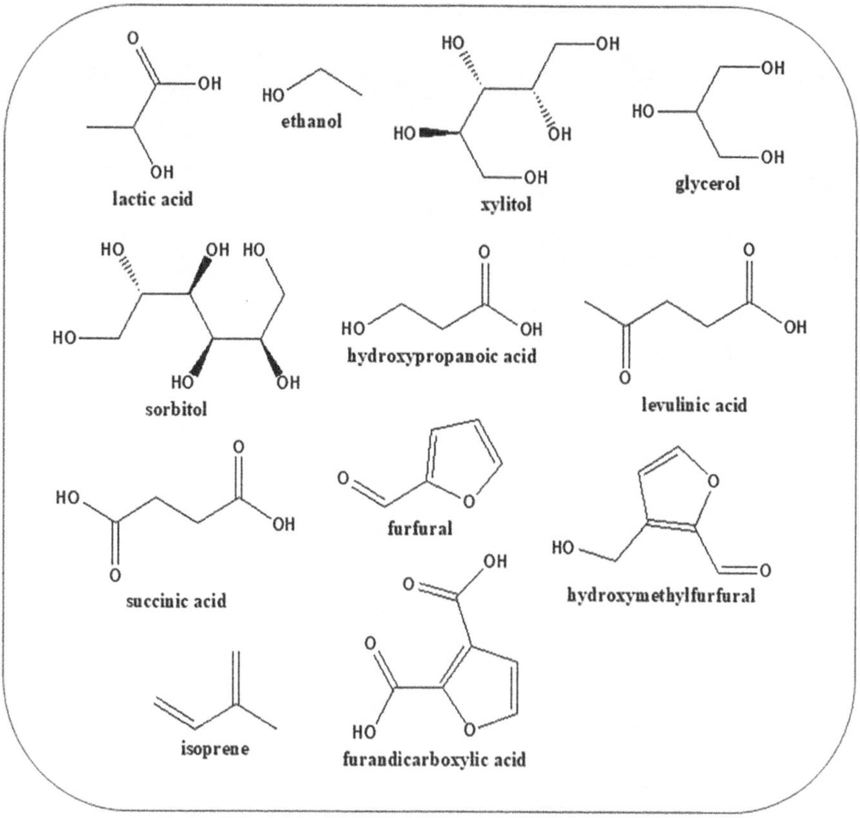

Fig. 1.3 Proposed biobased platform molecule

impact the overall market conditions and profitability (Cardona-Alzate et al. 2020). Thus, with the several factors governing the prices of crude oil refineries, biorefineries have proven to be a great and economic alternative (Figs. 1.4 and 1.5).

There is a need for both the stabilization of biomass pricing and the advancement of technologies to advance and optimize the usage of alternative feedstocks globally. Moreover, biorefineries offset the carbon footprints of fossil fuels because using fossil fuels results in significant CO_2 emissions that are penalized by society (Menon and Rao 2012). The absorption of CO_2 during biomass growth is believed to provide a balance and contribute to the potential benefits of biomass consumption in mitigating climate change. The viability of the hydrocarbons sector was determined by the market development for petrochemicals (Shahbaz et al. 2021). Initially, only a small portion of crude oil was consumed, leading to significant environmental damage due to the leftover waste. However, at that time, environmental concerns were not a priority, and the increase in energy demand facilitated the growth of unsustainable practices (Schnaiberg et al. 2002). With the concurrent development of fractionation and conversion techniques, driven by the demand for novel

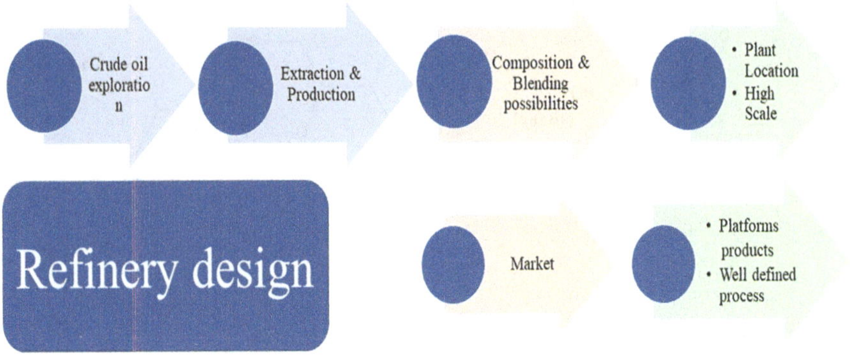

Fig. 1.4 Refinery design

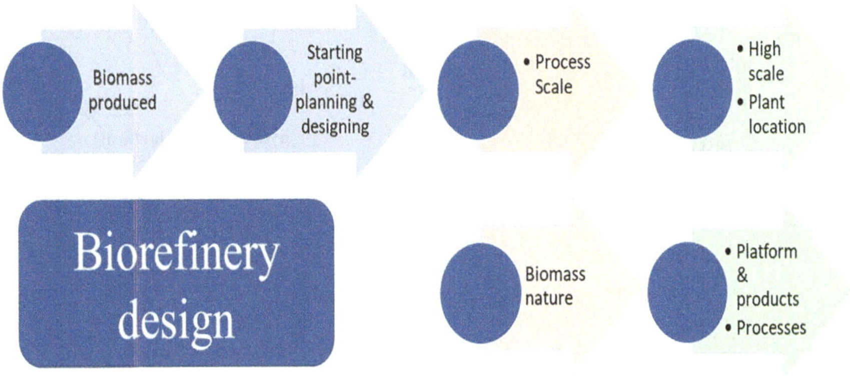

Fig. 1.5 Biorefinery design

products, especially plastics, the petrochemical industry gradually reduced the residues to minimal levels that are widely recognized today. This progress led to an increased reliance on crude oil as a primary raw material. Over time, crude oil gained extensive usage due to its abundance and suitability for the production of various petrochemical compounds. The industry's ability to efficiently process crude oil into different products has contributed to its widespread adoption as a key feedstock in petrochemical manufacturing. Comparing the software and tools available today with those from that era is particularly difficult due to the significantly lower quality and efficiency of the informatics systems back then (Zolfaghar et al. 2013). This method cannot be used using biomass as the basic material. The expansion of biorefineries needs to be examined in the new sustainability context that is now mandated. Although biomass is a complicated system and technology maturity isn't very high, it's still important to incorporate sustainability concepts and techniques, such as those in Table 1.2, during comprehensive design.

Table 1.2 Difference between the biorefineries and crude oil refineries based on sustainability principles as well as design strategies

S. No.	Refinery design strategy	Biorefinery design strategy	Comments
1	The same quality of crude oils is refined; rather specific crude oils are refined in a specifically designed refinery	Combining the feedstocks. The overall biomass efficiency and raw material utilization both rise as integration levels rise	The biorefinery approach should be maintained. New refineries now utilize different feedstocks such as natural and shale gas LEXINNOVA within the facility
2	The best method used in recent decades to produce petrochemicals, such as reactive distillation, was integrated technology	The most effective technologies combine fermentation, saccharification, and separation into one system	Biorefineries incorporated integrated petrochemical industry technologies
3	Refineries did not prioritize reducing waste streams as a primary goal. With time, design techniques such as pinch analysis were added	Based on various levels of integration, multiprocessing biorefineries combine various raw materials and products to produce additional products with added value	The design of biorefineries has always taken cues from the crude oil sector, such as the pinch analysis
4	Refineries were initially designed as independent facilities without conducting a comprehensive assessment of the feasible and practical products that could be obtained from them. The focus was primarily on the processing and refining of crude oil to produce conventional petroleum products such as gasoline, diesel, and jet fuel. At that time, the concept of integrated refining and petrochemical complexes was not extensively explored	The best scenario is always a multiproduct portfolio Technical, economic, environmental, and social process indicators typically rise as the number of goods increases	Any biomass-based project should take into account the maximum products that can be produced after using all biomass, based on preliminary or heuristic analysis
5	For many years, the refineries showed little concern for the ecosystems in general or the environment. The reputation of the crude oil sector today is poor mostly because of this well-known	The usage of other natural resources is decreased because of the preservation of ecosystems by second- and third-generation raw materials	Biorefineries must adopt the concept of life cycle analysis (LCA) and avoid making the same mistakes by giving environmental impact assessments the same weight as economic assessments
6	The most advanced and effective logistics are used in the oil business. The	The most widely used raw material in the world, biomass, is found in all	In comparison to oil refineries, the logistical and social aspects of

(continued)

Table 1.2 (continued)

S. No.	Refinery design strategy	Biorefinery design strategy	Comments
	actual societal impact is minimal	nations. Positive effects may result from effective logistics and a biorefinery, among other factors	biorefineries are more crucial. The supply chain will determine the project's true success given the massive amount of biomass that can be dispersed throughout many nations or regions
7	Energy usage and extremely complicated by-products and wastes were not a problem in crude oil refineries for a very long time	It is necessary to value by-products and reduce energy use during design	Energy consumption and the utilization of by-products are important to consider in the design of biorefineries when assessing the feasibility of a project
	The energy was regarded as being extremely abundant, and any leftover oil or by-products were simply dumped in open ponds or landfills		
8	The development of thermodynamics, kinetics, mass energy transfer, and other fields was fueled by the crude oil business. The most potent design software was created and developed by this sector from the start	Designing biorefineries with sustainability in mind is a crucial component. Using Aspen Plus and other specialized software is a thorough technique to examine the processes' convenient scale	Practically every design tool used in the crude oil business is used in the design of biorefineries. To be closer to the actual feedstocks and processes utilized in biomass conversion, specialized software is required
9	Based on cutting-edge approaches for reliable computations, crude oil refineries were extremely worried about safety even in the early stages	Create safer procedures. A good simulation aims to use chemicals and materials under controllable circumstances to guarantee the stability of the units	The design of a biorefinery takes safety into account using the same methods as the crude oil sector
10	The past, present, and future scenarios for oil refining (including allocation variables, market goods, etc.) are essentially never taken into account by the crude oil industry or petrochemical facilities	Sensitivity analysis could be incorporated into the design to better understand the potential uses of the biorefinery	Before implementation, scenarios and optimization techniques can reveal information about the biorefinery's true sustainability. This kind of examination is especially necessary for brand-new crude oil refineries because of the genuine risk that biomass poses to them
	The fact that we currently have a steady market for fuels and polymers, for instance, provides a good explanation for this		

(continued)

Table 1.2 (continued)

S. No.	Refinery design strategy	Biorefinery design strategy	Comments
11	To stabilize prices, the crude oil industry has implemented quota production models. This model remains functional even with varying levels of uncertainty	To prevent unanticipated and unfavorable fluctuations in the pricing of the energy and raw materials utilized in the biorefinery, hedging techniques should be taken into account. This enables the management of financial risks	In order to ensure stability in the business model regarding biomass prices, it might be necessary to establish an organization similar to the Organization of the Petroleum Exporting Countries (OPEC)

1.4 Classification of Biorefineries

Biorefineries are categorized based on the type of biomass utilized and the resulting products at each stage of biofuel production (de Jong and Jungmeier 2015). Cost-efficiency plays a crucial role in biofuel production, encompassing both transportation and manufacturing processes. The classification of biorefineries is determined by four main factors: the platform they operate on, the products generated, the feedstocks employed, and the conversion methods utilized (Cherubini et al. 2009b).

1.4.1 Biorefineries Can Be Classified in the Following Ways

1.4.1.1 Technology-Based Classification

This categorization includes conventional and advanced biorefineries, which are further divided into first-, second-, and third-generation technologies.

1.4.1.2 Raw Material-Based Classification

Biorefineries can be classified as whole crop biorefineries (WCBRs), oleochemical biorefineries, lignocellulosic feedstock biorefineries, green biorefineries, and marine biorefineries based on the type of raw materials they utilize.

1.4.1.3 By-product and Intermediate-Based Classification

This classification distinguish between the utilize syngas platforms and sugar platforms, depending upon the by-products.

1.4.1.4 Conversion Process-Based Classification

Biorefineries are further classified as thermochemical biorefineries, biochemical biorefineries, or two-platform concept biorefineries, based on the specific conversion processes employed.

Platforms refer to the intermediate substances generated during biofuel production, which can serve as final products or be utilized as by-products to connect different biorefineries. Examples of platforms include C_5/C_6 sugars, syngas, and biogas (de Jong et al. 2012). The complexity of the biorefinery system increases with the number of intermediates involved. Biorefineries mainly utilize two main feedstock groups: "energy crops" such as starch crops and rotation forestry, which are used to produce "energy products" like bioethanol, biodiesel, and synthetic biofuels. The other group is "biomass residues" such as straw, bark, wood chips, used cooking oils, and waste streams. These residues are used to produce various products including chemicals, materials, food, and feed (Cherubini et al. 2009a).

1.4.2 Examples of Biorefinery Classifications

The biorefineries can be classified into various forms on the basis of different parameters. If we classify them on the basis of material than four major biorefineries are as follows.

1.4.2.1 Oil Biorefinery

This type of biorefinery uses oilseed crops to produce biodiesel, glycerin, and animal feed.

1.4.2.2 C_6 Sugar Platform Biorefinery

These biorefineries produce bioethanol and animal feed using starch crops.

1.4.2.3 Syngas Platform Biorefinery

Utilizing straw, this type of biorefinery produces Fischer-Tropsch (FT) diesel and phenols.

1.4.2.4 Lignin Biorefinery

These biorefineries utilize wood chips to produce C_5 and C_6 sugars, electricity, heat, and phenols for bioethanol production.

The other way to classify the refineries, is on the basis of raw materials used.

1.4.3 Conventional Biorefineries

In 1839 a Dutch businessman Scholten started the industrial manufacture of potato starch which provided the base idea for the beginning of industrial refining which began in the nineteenth century with steam-powered paper machines (de Jong and Jungmeier 2015). In this process, high-quality native proteins were extracted from the potato juice using advanced methods, which improvise the quality of proteins making it suitable to use in food, cosmetics, and medicines as well as it can also be used as feedstock which makes it the best end food product. It also used in the manufacture of "Spiritus" or "Kartoffelsprit in German" which is the vodka or ethanol. Twenty-five percent of this product was incorporated with petrol until the 1950s. The Scholten company produced a variety of goods, including thermoplastic starch-based biopolymers, in addition to (modified) starch (Rahardiyan et al. 2023). Industrial biorefining was revolutionized by the advancement in food production like potato starch, wheat and corn starch, soy oil, proteins, vitamins, etc.

In 1801, Franz Achard of Germany introduced a series of processes including extraction, filtration, evaporation, crystallization, and centrifugation, which are used to produce crystalline sugar from beetroot (De Jong et al. 2010). Following World War II, soybeans gained popularity as a source of edible oil and as a substitute for protein-rich diets. Currently, soy is a major crop in countries like the USA, Brazil, Argentina, and Paraguay, with significant export volumes (Garrett et al. 2013). The process of obtaining soybean oil involves breaking the beans, controlling moisture levels, and rolling them into flakes. Soybeans are utilized in the production of various products such as margarine, butter, vegetarian burgers, oils, soap, cosmetics, inks, and clothes and increasingly as a feedstock for biodiesel (Fig. 1.6).

1.4.4 Advanced Biorefineries

An advanced biorefinery is a type of facility that shares similarities with conventional biorefineries as they both rely on a single feedstock. However, what sets advanced biorefineries apart is their capability to produce a diverse range of goods. Let's explore two specific examples of advanced biorefineries:

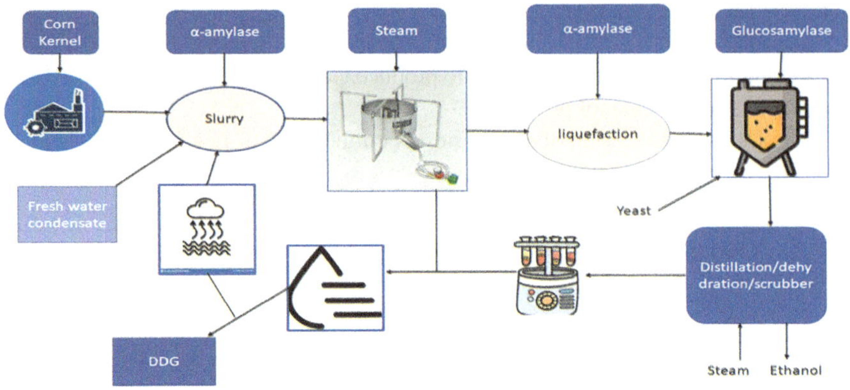

Fig. 1.6 Dry-mill ethanol process plant

1.4.4.1 Starch-Based Chemical Production

An advanced biorefinery of this kind utilizes starch as the primary feedstock. By employing advanced processing techniques, it can generate a wide variety of chemicals as end products. These chemicals find applications in various industries such as pharmaceuticals, textiles, and plastics. The advanced biorefinery's ability to convert starch into multiple valuable chemicals sets it apart from conventional biorefineries.

1.4.4.2 Cereal Grain-Based Carbohydrate Derivatives and Bioethanol Synthesis

Another example of an advanced biorefinery involves the use of cereal grains as the feedstock. Through the implementation of advanced processing methods, this biorefinery produces not only bioethanol but also a range of carbohydrate derivatives. These derivatives have applications in various industries like food, cosmetics, pharmaceuticals, etc. By efficiently converting cereal grains into both bioethanol and carbohydrate derivatives, this advanced biorefinery showcases its capability to produce multiple valuable products.

We can say that advanced biorefineries leverage advanced processing techniques to convert a single feedstock into various valuable products. These examples highlight the versatility and efficiency of advanced biorefineries in meeting the demands of different industries.

1.4.5 Whole Crop Biorefinery

In a WCBR, the fermentation process is used for the processing of grains and straw to obtain the final products, which is followed by dry and wet milling, and after distillation of grains and straw, these are converted into end products (de Jong and Jungmeier 2015). Sulfur dioxide is added to water-soaked grains during wet milling to lose the hull and soften the kernels. The starch, cellulose, oil, and proteins can then be separated from the grains using known processes (Rausch et al. 2019). Before the flour is blended with water and the enzymes are liquefied, whole grains, including the germ and bran, are ground in the dry milling procedure. The starch in the mash is then broken down by boiling it. This hydrolysis phase can be avoided by adding fermenting yeast and saccharifying enzymes to the fermenter at the same time. After fermentation, the alcohol in the mash is concentrated, purified, and dehydrated using a multicolumn distillation system (which is then turned into beer). The remaining mash (stillage), which is composed of both liquid (syrup) and solid (wet grains) phases, is combined, dried, and used as cow feed under the name "distiller's dried grains with solubles" (DDGS). The straw, consisting of chaff, nodes, ears, and leaves, is a lignocellulosic feedstock that can be further processed (Fernando et al. 2006) (Fig. 1.7).

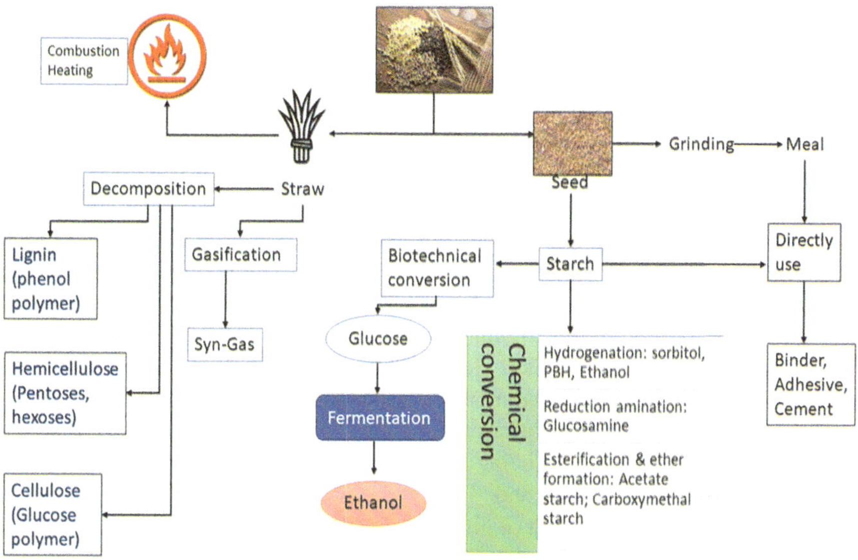

Fig. 1.7 Whole-crop biorefinery process

1.4.6 Oleochemical Biorefinery

An advanced version of Whole Crude Biodiesel Refinery (WCBR) involves the extraction of biodiesel while simultaneously producing high-value vegetable oil-based products. Advanced biorefineries utilize oil crop-derived fatty acids, esters, and glycerol to produce sustainable feedstocks, platform compounds, monomers, lubricants, and surfactants, aiding in the effective replacement of fossil fuel counterparts. An enormous quantity of lignocellulosic biowaste is used in lignocellulosic feedstock biorefineries. This biomass is also present in significant numbers in the precursor feedstocks used to make products based on vegetable oils (Schneider et al. 2016).

1.4.7 Lignocellulosic Feedstock Biorefinery

Lignocellulosic feedstock biorefinery is made up of three fundamental chemical fractions: (i) lignin, phenol polymers, (ii) cellulose, and (iii) hemicellulose. Hard fibrous plant materials obtained from municipal waste are used in LCF. To divide plant matter into its three components, it is first chemically cleaned and digested or hydrolyzed using enzymes (hemicellulose, cellulose, and lignin). Sulfite and alkaline (caustic soda) can produce cellulose and hemicellulose (acidic, bisulfite, alkaline, etc.). Plant-based lignin is broken down by enzymes such as ligninases, lignin peroxidases, laccases, and xylanolytic enzymes, leading to the hydrolysis of sugar components. Hemicellulose, a branched carbohydrate composed of five different sugars including uronic acid, C-5 sugars (D-galactose, D-glucose, D-mannose), and C-6 sugars (D-xylose, L-arabinose), is formed during this process. Xylose, a component of hemicellulose, is particularly valuable as it can be converted into furfural, a key building block for nylon and other applications. Hydrolyzing cellulose using chemical or enzymatic methods yields glucose and produces useful by-products such as ethanol, acetic acid, acetone, butanol, succinic acid, and other fermentation products (Calvo-Flores and Martin-Martinez 2022). While cellulose and hemicellulose have various applications, lignin is currently utilized primarily as an adhesive, binder, or direct combustion fuel. However, the lignin structure holds significant potential for producing additional monoaromatic hydrocarbons, which could greatly enhance the overall lignocellulosic fractionation (LCF) process if feasibly isolated. In Missouri, an LCF facility utilizing 4000 t of feedstock per day produces approximately 180,106 t of ethanol and 323,103 t of furfural annually. Additionally, significant microbial conversion of glucose can yield hydrogen, methane, propanol, and acetone, which are all petrochemical products (Calvo-Flores and Martin-Martinez 2022) (Fig. 1.8).

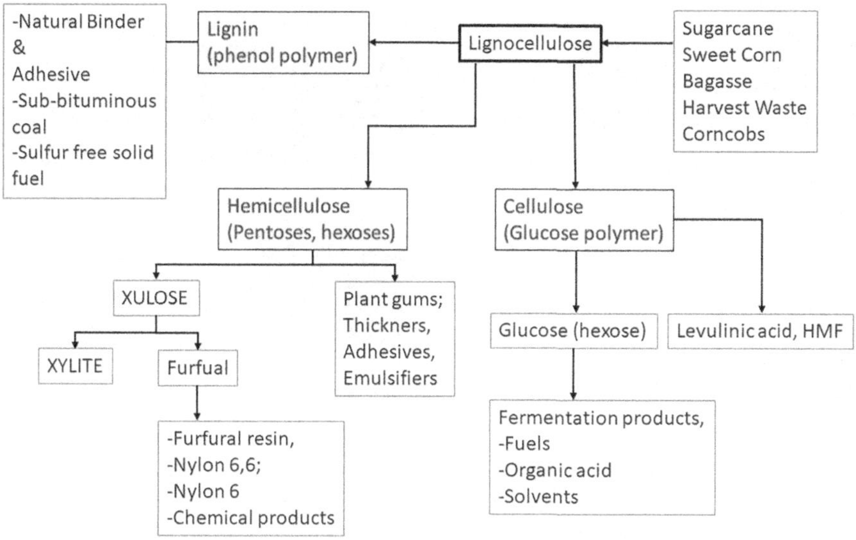

Fig. 1.8 LCF biorefinery process and products

1.4.8 Syngas Platform Biorefinery (Thermochemical Biorefinery)

The initial step in processing lignocellulosic biomass involves subjecting it to high temperature and pressure through size reduction, drying, or torrefaction methods. This treatment results in the production of gas, predominantly composed of carbon monoxide (CO) and hydrogen (H_2). The obtained syngas then undergoes purification in a high-temperature gas cleanup system, where impurities are removed. The composition of the gas is adjusted to meet the specific requirements of downstream synthesis processes.

Following this, a catalytic synthesis method is employed to extract various products from the clean gas. These products can include FT diesel, dimethylether, a range of alcohols, as well as several fundamental chemicals such as ethylene, propylene, butadiene, and others. The catalytic synthesis process enables the conversion of the syngas into these valuable end products, which find applications in industries such as transportation, chemical manufacturing, and energy production (Haro et al. 2013) (Fig. 1.9).

1.4.9 Next-Generation Hydrocarbon Biorefinery

Advancements in nanoscience and technology have introduced various new methods that aid in the development of fuel conversion techniques from biomass. However,

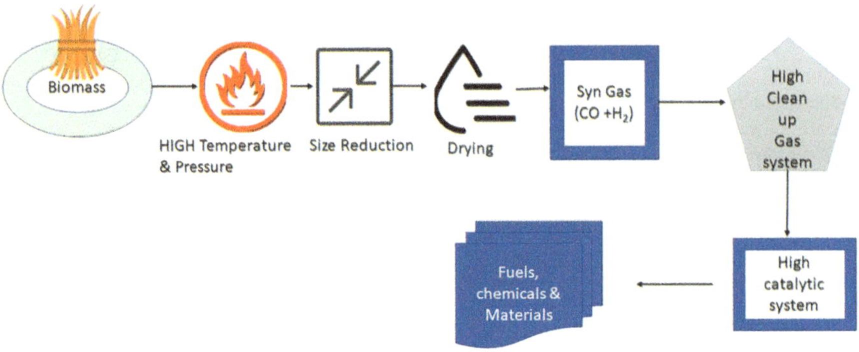

Fig. 1.9 Conceptual of syngas platform-based biorefinery

there remains a need to establish economically viable hydrocarbon biorefineries. One such method is liquid-phase catalytic processing, which involves the production of functionalized hydrocarbons from intermediates derived from biomass, such as hydroxymethylfurfural (HMF). This biorefining technique offers a means to convert biomass into valuable hydrocarbon products (Huber 2008).

Another method involves the utilization of renewable furan derivatives, which can serve as alternative raw materials for fossil fuels and polymers. These derivatives are derived from C_5 and C_6 carbohydrates found in biomass, such as sugars, hemicellulose, and cellulose. This approach enables the production of biofuels. Avantium Chemicals, a company based in the Netherlands, is currently involved in the development of chemical catalytic routes for this purpose. The development and implementation of these innovative methods and technologies are essential in advancing the field of biomass conversion and establishing sustainable and economically viable biorefineries.

1.4.10 *Green Biorefinery*

It was developed by a Dutch researcher that used grass and other leafy materials in the biorefineries with grass costing more per ton than leaves (€50–70). The mechanical refiner then degrades the leaf material to convert it into pure fiber with less than 11% protein content. The products thus formed are pressed to form juice, and the residual juice thus obtained is evaporated to obtain a concentrated product. This material is used as animal feed, building materials, insulation fiber, nursery pots, packaging material, feedstock for biofuels, etc. This range of greens used in biorefineries can be increased by including a wide variety of biomass. Occasionally, mixed feedstocks, including fresh and silage grass, can be employed in between lignocellulosic and green biorefineries. The majority of European green biorefinery initiatives now being worked on in Austria, Germany, Ireland, and the Netherlands,

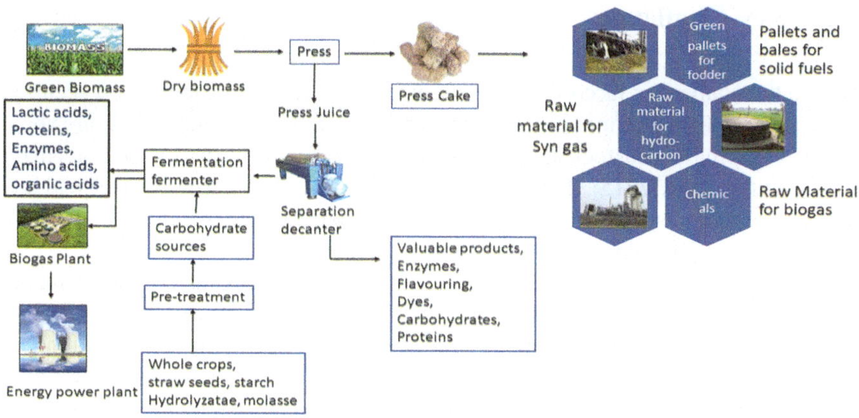

Fig. 1.10 Representation of green biorefinery process and products

center on grass refining. The objective is to extract useful materials with no waste and no emissions and to use all by-products in a biogas plant to make sure the plant can run on its energy (de Jong and Jungmeier 2015 n.d.) (Fig. 1.10).

1.4.11 Marine Biorefinery

Global biomass output is equally distributed between land and sea. Marine sources like microalgae and macroalgae offer potential, yet policies have mainly focused on terrestrial biomass. Diatoms, a type of microalgae, are the most common form of phytoplankton, with an estimated 100,000 species and significant biomass potential. Diatoms are known to accumulate oils and can vary in the production of oil, starch, sugar, and vitamins based on species and growing conditions.

Golden and green algae are capable of producing oils, starch, and carbohydrates; at the same time marine flora have been documented for their superior ability to absorb CO_2 compared to terrestrial species, making them a potential feedstock source for biofuels. However, the cost of biomass production and harvesting in marine biorefineries remains prohibitively expensive for widespread use in bulk chemical and fuel applications at present. Although marine biomass shows promise, further research and technological advancements are needed to make marine biorefineries economically viable and practical for large-scale chemical and fuel production (Antelo et al. 2015).

1.5 Key Challenges

Due to the challenges faced by the government like climate change, food security, and energy availability, more experiments and explorations are done to inquire about the set of biomass feedstock as an alternative for biofuels, bioenergy, and biochemical (Creutzig et al. 2015). These issues can be resolved by particular regional and national government initiatives. Yet, creating policies can be difficult because focusing on one issue may be detrimental to another.

In the absence of specific policies, it is impossible to establish a proper supply chain that ensures the availability of biomass feedstock. Therefore, there is a need to formulate legislation that governs the collection, transportation, and handling of biomass feedstock. Incentives should be provided to farmers to encourage their participation in delivering feedstock to next-generation ethanol plants. The involvement of farmers is crucial for the long-term sustainability of the ethanol industry, and the design of appropriate policies is essential to achieve this objective (Sharma et al. 2020). To mitigate risks to the ethanol industry, a continuous supply chain must be established. India, with its large population advantage, can capitalize on this by implementing special training programs like Skill India to address any skill gaps. Such initiatives will enhance the efficiency of the supply chain, maximize economic potential, and deliver social benefits. Moreover, they will contribute to environmental improvements and reduce the high market risks associated with next-generation ethanol industries.

The major obstacle in the bioethanol industry is the pretreatment of the feedstock. Pretreatment is required so that the carbohydrates present in the lignocellulose can be made available for conversion into the raw materials that are used in the cellulosic-based bioethanol industry. But this pretreatment makes the whole process highly cost-ineffective (Galbe and Zacchi 2007). This process is also important to avoid product degradation as to make the carbohydrates in the lignocellulose available for conversion is a significant obstacle in the manufacture of cellulosic-based bioethanol. In addition to costs, the most important factor in choosing a pretreatment procedure is its effectiveness in preventing product degradation which can prevent the future hydrolysis and enzymatic fermentation of the product (Tian et al. 2012).

1.5.1 Lignocellulosic Biorefinery Commercialization Difficulties

1.5.1.1 Economic and Technical Factors

It is a requirement for lignocellulose-based biorefineries to become industrialized to resolve several fundamental problems that arise at various stages of the processing of biomass. Atypical biomass conversion scheme consists of six main steps: (i) the choice of the biomass and its pre-processing; (ii) efficient pretreatment; (iii) the creation of a hydrolytic enzyme cocktail on-site or off-site; (iv) enzymatic saccharification; (v) hexose and pentose fermentation; and (vi) downstream processing (Tursi 2019).

1.5.1.2 Biodiesel Production

As biodiesel is made from biomass feedstock, food security is highly compromised during the process, and also practically it is impossible to use edible oil for bioethanol production as most of the edible oil supply in the country is imported. High cultivation costs, low production, and inadequate supply of raw materials also make the bioethanol production process cost-ineffective (Limayem and Ricke 2012). Furthermore, if nonedible oil is used instead of edible oil for bioethanol production, certain limitations arise. For instance, its performance may decrease in colder regions, and there is a possibility of contamination by animal fats or other impurities.

The process of bioethanol production has further challenges like the whole extraction process that needs further filtration which is plant specific and is unsuccessful on a large number of biomass feedstock, which makes the whole process highly expensive. Another limitation is that the production of n-hexane can cause severe effects on the environment as well as human health as it produces large amounts of intoxicants that will be added to water bodies as well as air in the form of volatile organic compound emissions. Thus, noticeable government support is required for the production of biodiesel, and this, the use of bioethanol, should be legalized and incentive support should also be provided by the government to promote the biodiesel program.

1.5.1.3 BioCNG Production

The main reason behind the failure of biogas plants is the reduction in the number of farm animals in rural areas which results in the non-availability of the raw materials used for the production of biogas. Any kind of fluctuation in the raw material supply can alter the plant's productivity as well as performance, and in turn, it results in the loss of the biogas plant.

Individual domestic biogas plants can only produce small-scale biogas which is used only as cooking fuel and a huge amount of monetary investment; however, the large-scale production of biogas is controlled by private industries which are working in partnership to make a huge profit by producing products like heat, power, or transportation fuel. Also, these days agriculture is more inclined toward the use of organic fertilizers in place of chemical fertilizers which reduces the availability of biomass for the production of biogas. In addition to the aforementioned factors, several other reasons contribute to the decline in biogas plants. These include inadequate feedstock supply, as well as a lack of awareness and acceptance regarding the wide range of biomass that can be utilized for biogas production. Certain biomass sources, such as human excreta and dead animal carcasses, are not widely accepted for biogas production. Therefore, it is essential to organize proper training and awareness programs in rural areas to highlight the various environmental benefits associated with biogas production. The lack of transportation and segregation of municipal solid waste in cities has hindered the rapid expansion of the biogas plant industry. Additionally, significant regional variations pose challenges in standardizing the process. Most processes used for treating organic and industrial waste, such as composting, vermicomposting, and waste-to-pellets, require transportation either from rural areas to cities or vice versa. This adds to the overall expenses and bulkiness of the process. In urban areas, industries tend to prefer inexpensive treatment methods like composting, which is a slow process. Furthermore, the cost of the procedure increases due to the additional infrastructure required for biogas cleaning.

1.5.2 Challenges in Third-Generation Biofuels: Algal Biofuel

Third-generation biofuels, such as algal biofuels, face significant threats and challenges that hinder their widespread commercial acceptance compared to other competing biomasses (Lee and Lavoie 2013). The expensive nature of algal biofuels is one major barrier, deterring industries from adopting them. Additionally, the process of algae converting atmospheric CO_2 to carbonic acid leads to a rise in the ecosystem's pH, impacting algal growth and hindering sunlight penetration in large algal blooms (Larsdotter 2006). In addition to these factors, several other reasons contribute to the difficulties faced by algal biofuels in gaining acceptance.

There is a lack of information regarding the effective use of algal biofuels in vehicles and equipment, with limited research in this field. Further exploration is necessary to establish their viability in various applications. Moreover, a high lipid content in the feedstock is crucial for efficient conversion of algae into biofuels. Lipids, which contain oils, can be used for production of biofuels from algae. Extracting and processing these lipids enables the generation of biofuels from algae, making it a valuable renewable energy source (Coma et al. 2017).

The process of converting algal blooms into biofuels is complex, involving multiple stages that make it difficult to implement and adopt (Coma et al. 2017). Another challenge is the high demand for fertilizers to create algal blooms, which consumes significant energy and adversely impacts the environment by increasing CO_2 levels at a rapid rate (Mata et al. 2010). Minimizing algal death caused by pests and pathogens is a significant challenge in algal monocultures. Lack of biodiversity makes them more susceptible to infestations and disease outbreaks, leading to reduced productivity and economic losses. Developing sustainable methods for protecting algal crops from pests and pathogens is crucial for the long-term viability of algal production systems (Smith and Crews 2014). Water demand poses a significant challenge as algae require substantial amounts of water to grow. However, rising temperatures due to global warming accelerate water evaporation, reducing water sources for large-scale algal bloom production. Algal biofuel production is expensive compared to fossil fuels and other biomass fuels, making it economically uncompetitive (Azar et al. 2006). Quality issues arise as not all species of algal blooms produce the same quantity and quality of oil, posing challenges for large-scale cultivation (Benemann 1992). Meeting the nutrient demands of algae, including CO_2, iron, phosphorus, sulfate, and nitrogen, along with water and light, presents significant challenges in the current scenario (Ghernaout and Ghernaout 2012). Overall, these challenges collectively hinder the commercial acceptance and widespread adoption of algal biofuels, necessitating further research, technological advancements, and cost-effective solutions to make them a viable and sustainable alternative energy source.

1.6 Conclusion

As the world faces increasing consumption of non-renewable fossil fuels, the need for alternative and environmentally friendly energy sources has become crucial. Biorefineries have emerged as a sustainable solution to produce marketable biobased products, such as petro-based refineries, promoting a greener economy. However, the adoption of biorefineries still faces challenges. Economic efficiency is a key obstacle, requiring researchers to develop new reactions, technologies, and carbon pathways for economically viable synthesis of substances. Additionally, the lignin platform, despite being abundant in nature, lacks effective chemical and biotechnological processes. Scaling up the production of aromatic compounds from lignin is a significant challenge for the next decade. The entire biorefinery production cycle is complex and poses technical, economic, ecological, sociological, and long-term challenges. The dominance of the fossil-based economy in the current market, along with factors like feedstock availability, market demand, resource recovery effectiveness, and sustainability, further complicates the situation. Conducting life cycle assessments can provide valuable insights into the environmental impact of biorefinery projects and guide process optimization.

Biorefineries must optimize biomass utilization to align with market expectations and compete cost-effectively with fossil fuels. Currently, petroleum refineries allocate a large majority of their output to fuel production, with only a small portion dedicated to organic compound production for the petrochemical industry. Biorefineries should strive to replicate this proportion by producing both fuels and organic compounds to meet market demands effectively. To achieve sustainability, biorefineries must reduce their carbon footprint through the use of renewable energy sources. This requires collaboration among stakeholders and integration with other sectors and relevant technologies. By addressing these challenges proactively, biorefineries can enhance sustainability, competitiveness, and successful integration within the broader energy landscape.

References

Alalwan HA, Alminshid AH, Aljaafari HAS (2019) Promising evolution of biofuel generations. Subject review. Renew Energy Focus 28:127–139. https://doi.org/10.1016/j.ref.2018.12.006

Alfaqiri A, Hossain NUI, Jaradat R, Abutabenjeh S, Keating CB, Khasawneh MT, Pinto CA (2019) A systemic approach for disruption risk assessment in oil and gas supply chains. Int J Crit Infrastruct 15(3):230–259

Antelo LT, de Hijas-Liste GM, Franco-Uría A, Alonso AA, Pérez-Martín RI (2015) Optimisation of processing routes for a marine biorefinery. J Clean Prod 104:489–501

Awogbemi O, Kallon DVV, Onuh EI, Aigbodion VS (2021) An overview of the classification, production and utilization of biofuels for internal combustion engine applications. Energies 14(18):5687

Azar C, Lindgren K, Larson E, Möllersten K (2006) Carbon capture and storage from fossil fuels and biomass–costs and potential role in stabilizing the atmosphere. Clim Chang 74(1–3):47–79

Ballantyne AP, Alden CB, Miller JB, Tans PP, White JWC (2012) Increase in observed net carbon dioxide uptake by land and oceans during the past 50 years. Nature 488(7409):70–72

Becker J, Lange A, Fabarius J, Wittmann C (2015) Top value platform chemicals: bio-based production of organic acids. Curr Opin Biotechnol 36:168–175

Béligon V, Noblecourt A, Christophe G, Lebert A, Larroche C, Fontanille P (2018) Proof of concept for biorefinery approach aiming at two bioenergy production compartments, hydrogen and biodiesel, coupled by an external membrane. Biofuels 9(2):163–174

Benemann JR (1992) Microalgae aquaculture feeds. J Appl Phycol 4:233–245

Berntsson T, Axegaard P, Backlund B, Samuelsson A, Berglin N, Lindgren K (2008) Swedish pulp mill biorefineries: a vision of future possibilities. The Swedish Energy Agency, Eskilstuna, Sodermanlands Lan, Sweden

Bhan C, Verma L, Singh J (2020) Alternative fuels for sustainable development. In: Shukla V, Kumar N (eds) Environmental concerns and sustainable development, Air, water and energy resources, vol 1. Springer, Singapore, pp 317–331

Bolan N, Sarkar B, Vithanage M, Singh G, Tsang DC, Mukhopadhyay R, Ramadass K, Vinu A, Sun Y, Ramanayaka S (2021) Distribution, behaviour, bioavailability and remediation of poly- and per-fluoroalkyl substances (PFAS) in solid biowastes and biowaste-treated soil. Environ Int 155:106600

Borras SM Jr, Franco JC, Isakson SR, Levidow L, Vervest P (2016) The rise of flex crops and commodities: implications for research. J Peasant Stud 43(1):93–115

Calvo-Flores FG, Martin-Martinez FJ (2022) Biorefineries: achievements and challenges for a bio-based economy. Front Chem 10:973417

Cardona-Alzate CA, Serna-Loaiza S, Ortiz-Sanchez M (2020) Sustainable biorefineries: what was learned from the design, analysis and implementation. J Sustain Dev Energy Water Environ Syst 8(1):88–117

Chen W-H, Lin B-J, Lin Y-Y, Chu Y-S, Ubando AT, Show PL, Ong HC, Chang J-S, Ho S-H, Culaba AB, Pétrissans A, Pétrissans M (2021) Progress in biomass torrefaction: principles, applications and challenges. Prog Energy Combust Sci 82:100887. https://doi.org/10.1016/j.pecs.2020.100887

Cherubini F (2010) The biorefinery concept: using biomass instead of oil for producing energy and chemicals. Energy Convers Manag 51(7):1412–1421

Cherubini F, Bird ND, Cowie A, Jungmeier G, Schlamadinger B, Woess-Gallasch S (2009a) Energy-and greenhouse gas-based LCA of biofuel and bioenergy systems: key issues, ranges and recommendations. Resour Conserv Recycl 53(8):434–447

Cherubini F, Jungmeier G, Wellisch M, Willke T, Skiadas I, Van Ree R, de Jong E (2009b) Toward a common classification approach for biorefinery systems. Biofuels Bioprod Biorefin 3(5): 534–546

Coma M, Martinez-Hernandez E, Abeln F, Raikova S, Donnelly J, Arnot TC, Allen MJ, Hong DD, Chuck CJ (2017) Organic waste as a sustainable feedstock for platform chemicals. Faraday Discuss 202:175–195

Creutzig F, Ravindranath NH, Berndes G, Bolwig S, Bright R, Cherubini F, Chum H, Corbera E, Delucchi M, Faaij A (2015) Bioenergy and climate change mitigation: an assessment. GCB Bioenergy 7(5):916–944

de Jong E, Jungmeier G (2015) Biorefinery concepts in comparison to petrochemical refineries. In: Pandey A, Hofer R, Taherzadeh M, Nampoothiri M, Larroche C (eds) Industrial biorefineries & white biotechnology. Elsevier, Amsterdam, pp 3–33

De Jong E, Van Ree R, Sanders JPM, Langeveld JWA (2010) Biorefineries: giving value to sustainable biomass use. In: Langeveld H, Sanders J, Meeusen M (eds) The biobased economy. Routledge, London, pp 132–151

de Jong E, Higson A, Walsh P, Wellisch M (2012) Bio-based chemicals value added products from biorefineries. IEA Bioenergy, Task42 Biorefinery 34:1–33

Demirbas A (2008) Biofuels sources, biofuel policy, biofuel economy and global biofuel projections. Energy Convers Manag 49(8):2106–2116

DwiPrasetyo W, Putra ZA, Bilad MR, Mahlia TMI, Wibisono Y, Nordin NAH, Wirzal MDH (2020) Insight into the sustainable integration of bio-and petroleum refineries for the production of fuels and chemicals. Polymers 12(5):1091

Fahd S, Fiorentino G, Mellino S, Ulgiati S (2012) Cropping bioenergy and biomaterials in marginal land: the added value of the biorefinery concept. Energy 37(1):79–93

Fernando S, Adhikari S, Chandrapal C, Murali N (2006) Biorefineries: current status, challenges, and future direction. Energy Fuel 20(4):1727–1737

Ferreira JA, Agnihotri S, Taherzadeh MJ (2019) Waste biorefinery. In: Taherzadeh M, Bolton K, Wong JWC, Pandey A (eds) Sustainable resource recovery and zero waste approaches. Elsevier, St. Louis, MO, pp 35–52

Galbe M, Zacchi G (2007) Pretreatment of lignocellulosic materials for efficient bioethanol production. Adv Biochem Eng Biotechnol 108:41–65

Garrett RD, Rueda X, Lambin EF (2013) Globalization's unexpected impact on soybean production in South America: linkages between preferences for non-genetically modified crops, eco-certifications, and land use. Environ Res Lett 8(4):044055

Ghernaout D, Ghernaout B (2012) On the concept of the future drinking water treatment plant: algae harvesting from the algal biomass for biodiesel production—a review. Desalin Water Treat 49(1–3):1–18

Gielen D, Boshell F, Saygin D, Bazilian MD, Wagner N, Gorini R (2019) The role of renewable energy in the global energy transformation. Energ Strat Rev 24:38–50

Gravitis J (1998) A biochemical approach to attributing value to biodiversity—the concept of the zero emissions biorefinery. In: Proceedings of the 4th annual world congress on zero emissions, Windhoek, Namibia

Groom MJ, Gray EM, Townsend PA (2008) Biofuels and biodiversity: principles for creating better policies for biofuel production. Conserv Biol 22(3):602–609

Haro P, Ollero P, Villanueva Perales AL, Vidal-Barrero F (2013) Potential routes for thermochemical biorefineries. Biofuels Bioprod Biorefin 7(5):551–572

Ho DP, Ngo HH, Guo W (2014) A mini review on renewable sources for biofuel. Bioresour Technol 169:742–749

Huber GW (2008) Breaking the chemical and engineering barriers to lignocellulosic biofuels: next generation hydrocarbon biorefineries. National Science Foundation, Chemical, Bioengineering, Environmental and Transport Systems Division, Washington, DC

Inyang V, Laseinde OT, Kanakana GM (2022) Techniques and applications of lignocellulose biomass sources as transport fuels and other bioproducts. Int J Low Carbon Technol 17:900–909

Jungmeier G (2014) The biorefinery fact sheet. IEA Bioenergy Task 42:1–48

Karigar CS, Rao SS (2011) Role of microbial enzymes in the bioremediation of pollutants: a review. Enzyme Res 2011:805187

Khatiwada B, Sunna A, Nevalainen H (2020) Molecular tools and applications of Euglena gracilis: from biorefineries to bioremediation. Biotechnol Bioeng 117(12):3952–3967
Kjärstad J, Johnsson F (2009) Resources and future supply of oil. Energy Policy 37(2):441–464
Kurian JK, Nair GR, Hussain A, Raghavan GV (2013) Feedstocks, logistics and pre-treatment processes for sustainable lignocellulosicbiorefineries: a comprehensive review. Renew Sust Energ Rev 25:205–219
Kumar S, Kalra S, Sahni T, Sidhu SK, Sarao LK (2023) Agricultural lignocellulosic waste to biofuels. In: Srivastava N, Verma B, Mishra P (eds) Agroindustrial waste for green fuel application. Clean energy production technologies, Springer, Singapore, pp 205–247. ISBN: 978-981-19-6229-5. https://doi.org/10.1007/978-981-19-6230-1_7
Lange J-P (2007) Lignocellulose conversion: an introduction to chemistry, process and economics. Biofuel Bioprod Biorefin 1(1):39–48
Larsdotter K (2006) Wastewater treatment with microalgae—a literature review. Vatten 62(1):31
Lee RA, Lavoie J-M (2013) From first-to third-generation biofuels: challenges of producing a commodity from a biomass of increasing complexity. Anim Front 3(2):6–11
Limayem A, Ricke SC (2012) Lignocellulosic biomass for bioethanol production: current perspectives, potential issues and future prospects. Prog Energy Combust Sci 38(4):449–467
Mata TM, Martins AA, Caetano NS (2010) Microalgae for biodiesel production and other applications: a review. Renew Sust Energ Rev 14(1):217–232
McCollum D, Yang C (2009) Achieving deep reductions in US transport greenhouse gas emissions: scenario analysis and policy implications. Energy Policy 37(12):5580–5596
Menon V, Rao M (2012) Trends in bioconversion of lignocellulose: biofuels, platform chemicals & biorefinery concept. Prog Energy Combust Sci 38(4):522–550
Misra V, Pandey SD (2005) Hazardous waste, impact on health and environment for development of better waste management strategies in future in India. Environ Int 31(3):417–431
Mohan SR (2016) Strategy and design of innovation policy road mapping for a waste biorefinery. Bioresour Technol 215:76–83
Nanda S, Azargohar R, Dalai AK, Kozinski JA (2015) An assessment on the sustainability of lignocellulosic biomass for biorefining. Renew Sust Energ Rev 50:925–941
Panda AK, Singh RK, Mishra DK (2010) Thermolysis of waste plastics to liquid fuel: a suitable method for plastic waste management and manufacture of value added products—a world prospective. Renew Sust Energ Rev 14(1):233–248
Rahardiyan D, Moko EM, Shun TJ, Keong LC (2023) Thermoplastic starch (TPS) bioplastic, the green solution for single-use petroleum plastic food packaging—a review. Enzyme Microb Technol 168:110260
Ramachandra TV, Aithal BH, Sreejith K (2015) GHG footprint of major cities in India. Renew Sust Energ Rev 44:473–495
Rausch KD, Hummel D, Johnson LA, May JB (2019) Wet milling: the basis for corn biorefineries. In: Serna-Salvidar SO (ed) Corn. Elsevier, Amsterdam, pp 501–535
Schnaiberg A, Pellow DN, Weinberg A (2002) The treadmill of production and the environmental state. Enviro State Under Pressure 10:15–32
Schneider MN, Iaconi A, Larocca S (2016) Oleochemical biorefinery. In: Cavani F, Albonetti S, Basile F, Gandini A (eds) Chemicals and fuels from bio-based building blocks. Wiley, Hoboken, pp 519–534
Seghetta M, Hou X, Bastianoni S, Bjerre A-B, Thomsen M (2016) Life cycle assessment of macroalgalbiorefinery for the production of ethanol, proteins and fertilizers–a step towards a regenerative bioeconomy. J Clean Prod 137:1158–1169
Shahbaz M, AlNouss A, Ghiat I, Mckay G, Mackey H, Elkhalifa S, Al-Ansari T (2021) A comprehensive review of biomass based thermochemical conversion technologies integrated with CO2 capture and utilisation within BECCS networks. Resour Conserv Recycl 173:105734
Sharma YC, Singh V (2017) Microalgal biodiesel: a possible solution for India's energy security. Renew Sust Energ Rev 67:72–88
Sharma B, Larroche C, Dussap C-G (2020) Comprehensive assessment of 2G bioethanol production. Bioresour Technol 313:123630
Sinclair TR, Weiss A (2010) Principles of ecology in plant production. CABI, Wallingford

Smith VH, Crews T (2014) Applying ecological principles of crop cultivation in large-scale algal biomass production. Algal Res 4:23–34

Spliethoff H (2010) Power generation from solid fuels. Springer Science & Business Media, Heidelberg

Stedile T, Ender L, Meier HF, Simionatto EL, Wiggers VR (2015) Comparison between physical properties and chemical composition of bio-oils derived from lignocellulose and triglyceride sources. Renew Sust Energ Rev 50:92–108

Stegmann P, Londo M, Junginger M (2020) The circular bioeconomy: its elements and role in European bioeconomy clusters. Resour Conserv Recycl X 6:100029

Suarez-Bertoa R, Zardini AA, Keuken H, Astorga C (2015) Impact of ethanol containing gasoline blends on emissions from a flex-fuel vehicle tested over the worldwide harmonized light duty test cycle (WLTC). Fuel 143:173–182

Sudha P, Jamali PV, Sudagar IP (2023) Emerging improvements in the characterization and advanced utilization of woodland fibres and its hybrid composites. Ricerca Int J Multidiscip Res Innov 2(10):1–49

Verma D, Sahni T, Kumar S, Sarao, LK (2023) Synthesis and Application of Nanoengineered Cellulosic Biomass in Biohydrogen Production. In: Srivastava M, Mishra PK (eds) NanoBioenergy: application and sustainability assessment. Clean energy production technologies, Springer, Singapore, pp 135–167. ISBN: 978-981-19-6234-9. https://doi.org/10.1007/978-981-19-6234-9_6

Takkellapati S, Li T, Gonzalez MA (2018) An overview of biorefinery-derived platform chemicals from a cellulose and hemicellulose biorefinery. Clean Techn Environ Policy 20:1615–1630

Thang VH, Novalin S (2008) Green biorefinery: separation of lactic acid from grass silage juice by chromatography using neutral polymeric resin. Bioresour Technol 99(10):4368–4379

Themelis NJ (2003) An overview of the global waste-to-energy industry. Waste Manage World:40–48

Tian X, Fang Z, Guo F (2012) Impact and prospective of fungal pre-treatment of lignocellulosic biomass for enzymatic hydrolysis. Biofuels Bioprod Biorefin 6(3):335–350

Tursi A (2019) A review on biomass: importance, chemistry, classification, and conversion. Biofuel Res J 6(2):962

Tye YY, Lee KT, Abdullah WNW, Leh CP (2016) The world availability of non-wood lignocellulosic biomass for the production of cellulosic ethanol and potential pretreatments for the enhancement of enzymatic saccharification. Renew Sust Energ Rev 60:155–172

Ullah K, Sharma VK, Dhingra S, Braccio G, Ahmad M, Sofia S (2015) Assessing the lignocellulosic biomass resources potential in developing countries: a critical review. Renew Sust Energ Rev 51:682–698

Wan G, Bei L, Yu J, Xu L, Sun L (2022) Products distribution and hazardous elements migration during pyrolysis of oily sludge from the oil refining process. Chemosphere 288:132524

Wang MQ, Han J, Haq Z, Tyner WE, Wu M, Elgowainy A (2011) Energy and greenhouse gas emission effects of corn and cellulosic ethanol with technology improvements and land use changes. Biomass Bioenergy 35(5):1885–1896

Winsley P (2007) Biochar and bioenergy production for climate change mitigation. N Z Sci Rev 64(1):5–10

Yusuf N, Kamarudin SK, Yaakub Z (2011) Overview on the current trends in biodiesel production. Energy Convers Manag 52(7):2741–2751

Zolfaghar K, Meadem N, Teredesai A, Roy SB, Chin S-C, Muckian B (2013) Big data solutions for predicting risk-of-readmission for congestive heart failure patients. 2013 IEEE international conference on big data, Santa Clara, Silicon Valley, CA, USA 64–71

Chapter 2
From Organic Waste to Renewable Energy: Rice Straw

Bindu Bala and Sachin Kumar

Abstract Organic waste production poses significant global challenges, with mismanagement leading to adverse consequences for individuals and the environment. Anaerobic digestion, a biological treatment technique, offers a solution for processing diverse biological wastes while also producing renewable energy. This chapter explores the impact of various parameters on generation of biogas during digestion of rice straw anaerobically. Approaches like codigestion with various substrates are found effective in enhancing nutritional balance and increasing biomethane outputs. Different pretreatment techniques are discussed as means to improve digester performance. The review also addresses the digestion of rice straw on various scales, considering strategies to overcome barriers and enhance biomethane production. By understanding the factors influencing anaerobic digestion and employing innovative tactics, this research aims to contribute to sustainable waste management practices by optimizing biogas generation and energy production while mitigating environmental impacts.

Keywords Rice straw · Waste · Anaerobic digestion · Biogas production · Agriculture

2.1 Introduction

Half of the globe's population relies primarily on rice as a source of calories and nourishment (Arvanitoyannis and Tserkezou 2008), but its waste, i.e., rice straw (RS), is a big problem for environmentalists. Various researchers are working to convert this waste into a reliable and renewable energy source for the world (Zhao et al. 2010). After the grain is harvested, RS is a fibrous, lignocellulosic residue that stays in the field. When microorganisms break down lignocellulosic substrate under

B. Bala
Punjab Agricultural University, Ludhiana, Punjab, India

S. Kumar (✉)
Food Corporation of India, Ludhiana, Punjab, India

N. Srivastava et al. (eds.), *Paddy Straw Waste for Biorefinery Applications*, Clean Energy Production Technologies, https://doi.org/10.1007/978-981-99-8224-0_2

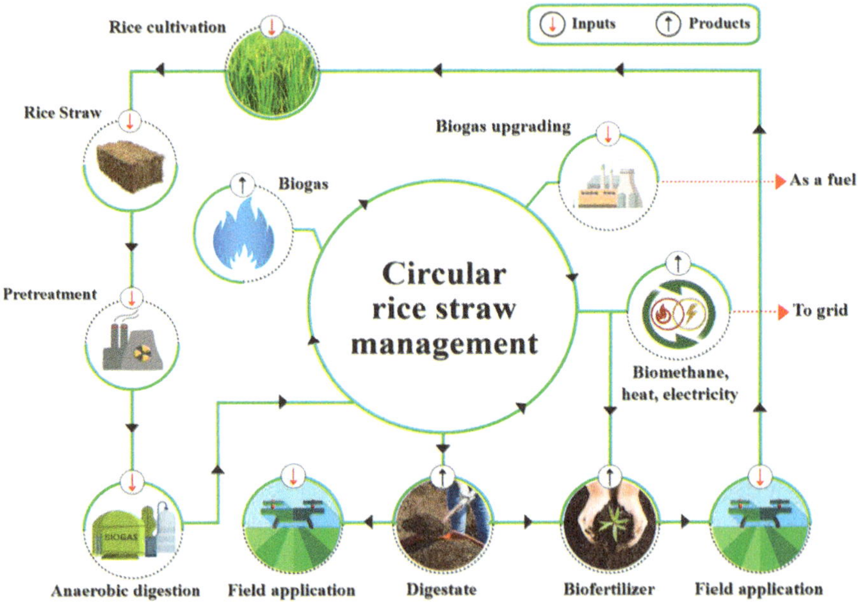

Fig. 2.1 Rice straw management (Source: Alengebawy et al. 2023)

anaerobic circumstances, they produce biogas that is rich in nutrients and carbon dioxide. This process is termed as "biomethanation" (Pore et al. 2016). Among other gas compositions, the biogas mixture primarily consists of methane, carbon dioxide, oxygen, nitrogen, and hydrogen sulfide. Since CH_4 has the greatest amount of energy of all the biogas constituents, it is also the most important component of biogas. Methane cannot be further reduced or oxidized, making it the ultimate reduced form and carbon dioxide the ultimate oxidized form in the anaerobic digestion (AD) of organic substrate (Fig. 2.1).

The amount of organic carbon that is reduced determines how much biomethane is produced during the process (Rajagopal et al. 2013). As more substrates are made accessible for their use, the need for anaerobic conversion of lignocellulosic wastes is growing (Riya et al. 2018). RS has a significant potential to be transformed to biomethane among the range of agricultural waste products that are currently available (Pore et al. 2016). It is one of the widely used lignocellulosic substance in bioenergy liberation worldwide (Mustafa et al. 2016). Using agricultural straw and manure as fuel, AD is a practical way to produce renewable energy (Forster-Carneiro et al. 2008). This chapter is especially dedicated toward the process and mechanisms involved to convert the RS into various useful products with special with special emphasis on AD.

2.2 Process of AD

2.2.1 Feedstock

Various types of feedstocks can be treated using AD. Yet, biomass made of lignocellulose provides a viable source of alternative energy due to its significant potential for the production of biogas (Hu and Ragauskas 2012). This kind of waste material, which includes agricultural residues (such as crop residues), wood, and grass, is widely accessible. Due to the intricacy of these materials' structures, which contain between 10% and 25% lignin, process repression is frequently observed when these elements are digested (Sawatdeenarunat et al. 2015). As a result, pretreatment is frequently required before using AD to produce biogas, while it can also be codigested along with other feedstock (Yang et al. 2015). For such feedstock, solid-state anaerobic digestion (ss-AD) digesters are the best option (Zhou and Wen 2019).

2.2.2 Feedstock Degradation Pathway

AD is required for the biogas production process, which produces renewable energy and nutrients from waste products. Feedstock degradation involves several unique and sequential phases or steps.

2.2.2.1 Hydrolysis

The majority of biological substrates are macromolecules like lipids, proteins, and polysaccharides. When these macromolecules are absent, monosaccharides are first produced. The existence and ability of bacteria like *Clostridium* or *Bacillus* to release extracellular enzymes determine the effectiveness of this stage. The occurrence of lignin in lignocellulosic substrates, with hemicellulose and cellulose units, creates a 3D molecule. Therefore, hydrolysis is recognized as a phase that significantly restricts the general rate of degradation, especially for these types of substrates (Sträuber et al. 2012). The biomass is shielded from the enzymatic attack by this physical barrier. Therefore, higher enzyme activity in bacteria is necessary for effective breakdown, especially when dealing with lignocellulosic biomass (Tsapekos 2017) (Fig. 2.2).

2.2.2.2 Acidogenesis

The next step is acidogenesis, and it involves fermenting the hydrolyzed by-products of macromolecules in accordance with several metabolic pathways. Various

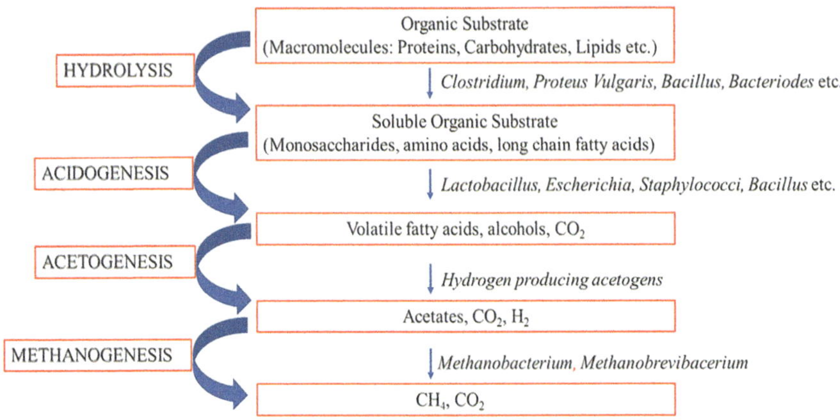

Fig. 2.2 Four major phases of feedstock degradation pathway

facultative and obligate bacteria produce a variety of volatile fatty acids (VFA), H_2, CO_2, and alcohols (*Staphylococcus*). Primary substrates for this process are sugar and amino acids. Propionate and biomass are produced as a result of glycerol fermentation (Angelidaki et al. 2011; Tsapekos 2017). High hydrogen concentrations produced by acidogenesis bacteria during this phase may hinder the synthesis of acetate.

2.2.2.3 Acetogenesis

In acetogenesis, longer fatty acids get transformed into H_2, CO_2, and acetates. Acetogens, which produce hydrogen, employ the products of acidogenesis as electron receivers and use the H_2 and CO_2 ions as electron donors. Maintaining a syntrophic relationship with methanogens is crucial to ensuring that acetogenic events take place in an energetically favorable fashion because this biological conversion route is not exergonic (Treu et al. 2016; Tsapekos 2017).

2.2.2.4 Methanogenesis

Methanobacterium, *Methanobrevibacerium*, and *Methanoplanus*, which are acetotrophic and hydrogenotrophic methanogens, work to convert acetate, H_2, and CO_2 into methane and CO_2 in this last stage. Methane is mostly created from H_2 and CO_2, with a smaller amount coming from the conversion of acetate (Angelidaki et al. 2011). Alternately, based on the process variables, H_2 and CO_2 could be created through a combination of the hydrogenitrophic pathway and synthrophic acetate oxidation (Campanaro et al. 2016).

2.3 Factors Affecting AD

2.3.1 Temperature

The effectiveness of process is significantly directed by temperature in terms of reaction velocity, physical diffusion, and chemical breakdown (Atelge et al. 2020). The fundamental requirement for lowering the susceptibility of the anaerobic system and ensuring its durability is optimal temperature (Shetty et al. 2017). A typical temperature range for AD bacteria is between 10 and 20 °C (psychrophilic), 30 and 40 °C (mesophilic), and 50 and 60 °C (thermophilic). Mesophilic and thermophilic temperature ranges are ideal for the microbe's rate of development because they also deactivate most of the population which is harmful (Takdastan et al. 2005; Hagos et al. 2017). The thermophilic range has advantages over the mesophilic range, like early degradation and a higher rate of organic loading, which leads to increased efficiency, but generation of FA results in acidification. Biogas' production is hampered by this acidity. Further drawbacks of AD in the thermophilic range include decreased stability, poor effluent quality, increased toxicity, and reduced methanogenesis. Mesophilic AD has stronger process stability and better diversity of anaerobic bacteria, although they release less biogas than thermophilic AD. Hence, methanogenesis in the mesophilic region and hydrolysis in the thermophilic range would be the ideal circumstances for AD (Ward et al. 2008) (Fig. 2.3).

2.3.2 VFA and pH

The range of pH between 6.6 and 7.6 is ideal for the biological process of AD (McCarty 1964). The ideal pH range for AD, according to Sreekrishnan et al. (2004) and Gerardi (2003), is 6.81–7.22. While bacteria producing methane are hampered below 5.5, acidogenic bacteria continue to be active. Low pH conditions are avoided because methanogenic bacteria are inhibited, lowering the amount of methane in the biogas. If there is a significant buildup of fatty acids during the acidogenesis stage, pH will often decrease during the AD process and fall below 6.6. Due to the rapid

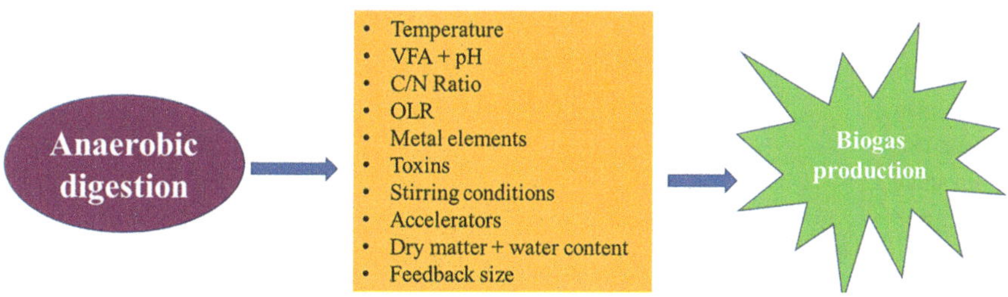

Fig. 2.3 Factors affecting AD of RS

metabolism of hydrogenic, acidogenic, and acetogenic bacteria, VFAs can build up. When they do, a significant amount of them can cause the reactor to foul by increase in the hydrogen ions (Luo et al. 2018).

The toxic substances in the feedstock prevent the methanogens from producing methane, which leads to pH fall, or the substrates becoming overloaded. In this situation, it is necessary to stop substrate feeding in order to stop or reduce acid production and allow acetogens and methanogens to breakdown any extra acid that was created. Using lime to neutralize the acid and raise the pH to a desirable level is another option. Moreover, AD is inhibited by a rise in pH (higher than 8.0). Methanogenesis fully ends when the pH reaches 9.0 (Clark and Speece 1971).

In the reactor sodium hydroxide and $NaHCO_3$ can be added to adjust pH when it turns sour (Ye et al. 2013). The ability of a solution to neutralize acid is known as alkalinity. The ions employed to boost the alkaline conditions in the digester include bicarbonate (HCO_3; CO_3^{2-} and OH).

2.3.3 *Carbon to Nitrogen Ratio*

Optimal bacterial growth achieved when the carbon to nitrogen (C/N) proportion falls between 20 and 30. When the C/N ratio surpasses this ideal range, the breakdown process slows down. Total ammonia nitrogen (TAN) and NH_3 may be produced during AD of a nitrogen-rich substrate, which, when combined with a high temperature and pH, may impede synthesis of biogas (Hansen et al. 1998). Unbalanced C/N proportion can prevent high ammonia nitrogen levels from occurring and cause VFAs to accumulate (Meng et al. 2018) (Table 2.1).

Also, it is vital to observe that organic wastes utilized to produce biogas are typically abundant in compounds resistant to lignin and cellulose (Yang et al. 2015).

Table 2.1 C/N ratio for different waste

Waste	C/N ratio	References
Kitchen waste	26:30	Guillaume and Lendormi (2015)
Food waste	2:18	
Sugarcane waste	139:151	
Vegetable waste	8:36	
Grass cutting waste	11:15	
Slaughterhouse waste	21:36	
Cattle manure	15:26	
Wheat straw	51:151	
Corn straw	51:57	
Sawdust	199:501	
Algae	74:101	
Sheep manure	20:34	
Chicken manure	4:16	

In such scenario, pretreatments are required to use such wastes with anaerobic organisms for brief retention times (Kabir et al. 2015). As a result, these materials include a sizable quantity of "invalid carbon," which has an impact on how C/N is calculated. Because of this, C/N calculations only depict the broad properties of organic wastes rather than the specific elements that anaerobic microorganisms really use.

2.3.4 Organic Loading Rate

The anaerobic digestion (AD) process is affected by the organic loading rate (OLR). It is crucial to consider this aspect due to the potential failure of the digester caused by overloading the feedstock, as well as the extensive research on the impact of OLR on digester performance (Nair 2013). While a high OLR can boost biogas generation rate, it lowers the system' stability. When organic loading rates reach their maximum, biomethane yield increases, but above that point, overloading causes a drop in both biomethane yield and the rate at which volatile solids degrade (Hashimoto 1986). The efficiency of the RS AD system might be improved by using a low organic loading rate and fewer feedings (Zealand et al. 2017). According to biogas plant tests, the digester achieved a volatile solid decrease of 66.67% while still producing the most gas at an OLR of 0.24 g/m^3 (Nagao et al. 2012).

2.3.5 Metal Elements

The successful operation of AD is definitely reliant on the presence of trace materials, and their deficiency can lead to disturbances in the reactor (Choong et al. 2016). During the organic matter breakdown, certain metal ions, including sodium (Na^{+}), potassium (K^{+}), calcium (Ca^{2+}), ferric (Fe^{3+}), and magnesium (Mg^{2+}), are produced. A higher concentration of these ions would inhibit microbial development or potentially be harmful since the osmotic pressure they produce causes bacterial cells to become dehydrated (Soto et al. 1993; Yerkes et al. 1997). When Mg^{2+} concentration reaches 720 mg/L, methanogenesis is observed to be inhibited. Moreover, methanogens may disaggregate at high magnesium ion concentrations (>100 mM), which prevents the conversion of acetate (Schmidt and Ahring 1993).

Doses above 300 mg/L, Ca^{2+} was modestly suppressive to bacteria. Nevertheless, high Ca^{2+} can result in carbonates and phosphates precipitating, which scales biomass, reactors, and pipes. This lowers the specific methanogenic activity and diminishes buffer capacity (Schmidt and Ahring 1993; Ariunbaatar et al. 2014). Zhang et al. (2009) reported that iron (Fe^{3+}) concentrations of 21 microgram per liter or higher can inhibit 50–85% of biogas production activities by deactivating enzymes within microorganisms through interactions with their functional groups. However, the production of methane from a high Zn^{2+} concentration is larger than that of the digester with no Zn^{2+}. On the other hand, adding Zn^{2+} at a slightly enhanced level of 200 mg/L hindered methane productivity (Cai et al. 2017).

2.3.6 *Toxins*

When a substrate is degraded, toxic chemicals are either produced or already exist in the system. Sulfate reducers and methanogens are both fatal to hydrogen sulfide (H_2S) at the non-dissociated concentration (Abudi et al. 2016). This form is thought to be the most dangerous because it may easily diffuse past a cell's membrane. Proteins are denatured as a result, and bacteria's assimilatory metabolism is also hampered. Sulfur and H_2S concentrations of less than 0.003 and 0.002 mole/L, respectively, are regarded as inhibiting.

According to several studies, the unionized content of sulfur dioxide in the 6.8–7.2 pH range is connected to toxicity (Anjum et al. 2017). Other hazardous materials that prevent microbial development in the reactor include detergents and minerals with heavy metal ions. Lower quantities of these compounds encourage the growth, while bigger concentrations may prevent it (Appels et al. 2008).

2.3.7 *Stirring Conditions*

Stirring plays a crucial role in maximizing the interaction between bacteria and organic waste, thereby accelerating the AD. Stirring also hinders the accumulation of solids at the digesters' base and inhibits the accumulation of foam and scum on the outermost layer of the fermentation solution (Fig. 2.4) (Haider et al. 2015). In the digester, the presence of "dead zones" characterized by low bacterial populations and inadequate breakdown of organic matter can hinder the total effectiveness of the process. Furthermore, some areas may experience the accumulation and settling of organic matter. To mitigate these issues, stirring or mixing is employed to promote the breakdown process. The stirring can be achieved through continuous or intermittent blending. While high-speed mixing has been found to reduce biogas production, moderate-speed mixing allows the digester to accommodate interruptions caused by shock loading.

2.3.8 *Accelerator*

Adding accelerants to the AD process increases production of biogas by facilitating the adsorption of microbes onto the substrate and enhancing the localized surface area. These accelerants can be categorized into organic additives, biological additives, and inorganic additives. Organic additives encompass a range of biomasses,

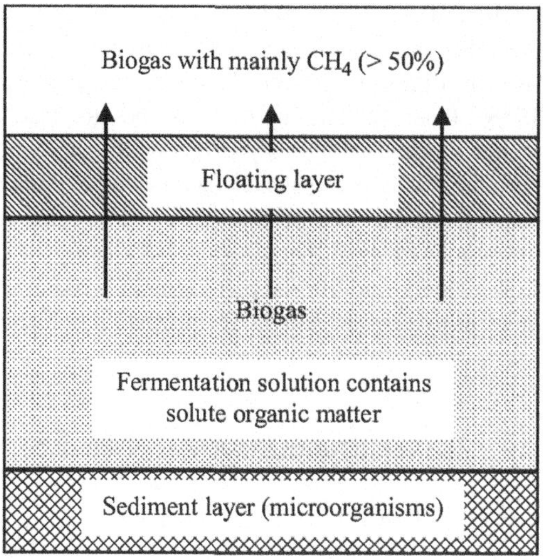

Fig. 2.4 Different layers in unstirred digester (Source: Ngan et al. 2020)

such as plant extracts, weed by-products, lignocellulosic biomass, and ensilage, which are derived from greenery sources. By incorporating these accelerants, the AD process can create more favorable conditions for microbial activity and enhance the efficiency of biogas production. The naturally occurring steroids in this extract function as microbial metabolic stimulators. In earlier experiments, stimulants like teresin and aquasan were employed to treat AD. The availability of cellulose and hemicellulose is increased by biological additives like fungal, microbial consortium, enzyme, etc., leading to improved digestibility.

2.3.9 *Water Content and Dry Matter*

For biogas generation and solid reduction, a digester's dry matter level should be between 9% and 11%. The amount of dry matter up to 20% aids in saving 45–55% of the digester capacity but may cause foul smell (decrease of pH), which in turn lowers the yield of methane. Monnet (2003) suggested adjusting the dry matter concentration of lot digesters to a range of between 5% and 10%.

2.3.10 *Feedback Size*

One of the elements influencing the generation of biogas is the size of the feedstock. Overly big materials should be avoided since they can clog digesters and hinder bacterial decomposition. Smaller materials will decompose more quickly because of their greater surface area and microbial activity.

According to several additional research, compared to digesters employing untreated substrates, physical pretreatment techniques can drastically minimize size of digester without lowering methane production (Gollakota and Meher 1988). It is challenging for microbes to decompose lignocellulosic materials on their own due to their high lignocellulosic content. Farmers often dispose of and burn the waste rice husks. In fact, it can be damaging to farmers and the environment because the burning process creates CO_2 gas, which is bad for your health. In the meantime, trash in the soil may be a cause of the disease in the subsequent crop. Consequently, additional pretreatment is required before fermentation in the digester in order to produce biogas from waste rice husk (Taherzadeh and Karimi 2008).

2.4 Pretreatment Method (Fig. 2.5)

2.4.1 Physical Pretreatment

This pretreatment may enhance the substrates' porosity and area of accessibility while decreasing the level of cellulose polymerization and crystallinity (Harmsen et al. 2010). Ball-milling wheat straw prior to use increased the specific surface area by 2.3 m^2/g as opposed to 0.64 m^2/g of raw straw (Gharpuray et al. 1983). Methane output and fiber degradation increase with particle size reduction from 90 to 2 mm (Mshandete et al. 2006). As a result, digestion time was shortened (23–59%) and methane output rose (5–25%) (Hendriks and Zeeman 2009). RS that has been chopped into 25-mm lengths contains more methane than RS that has not been cut (Zhang and Zhang 1999). However, because it uses a lot of energy, adding a milling phase to the AD process is expensive (Hendriks and Zeeman 2009).

Physical pretreatments are commonly employed in anaerobic digestion processes and can include techniques such as milling, extrusion, grinding, steam explosion, and liquid hot water treatment.

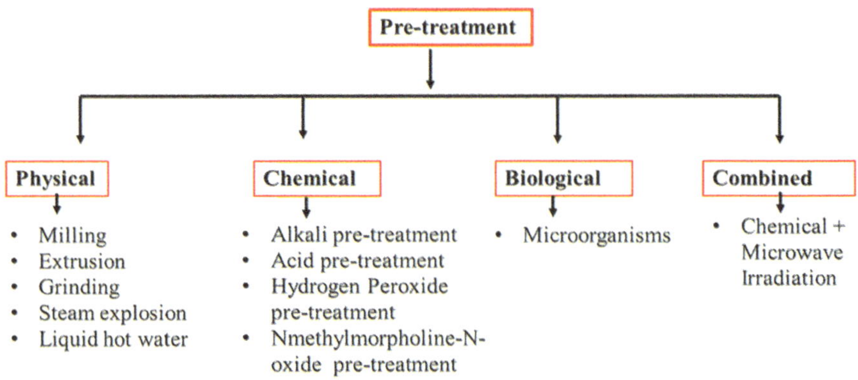

Fig. 2.5 Various pretreatment methods involved in AD

2.4.2 Chemical Pretreatment

The effectiveness of this process, which primarily relies on the breakdown or oxidation of molecules of organic matter utilizing acids, bases, and oxidants, is greatly influenced by the properties of the substrate (Salihu and Alam 2016).

2.4.2.1 Alkali Pretreatment

Alkaline pretreatment is more efficient than adding Co^{2+}, Ni^{2+}, and Se^{6+} and improved RS's biomethane productivity by 21.4% (Mancini et al. 2018). By destroying the ester and glucoside side chains, alkali pretreatment damaged lignin, increasing the substrate's porosity and making it more accessible to bacteria. The greatest outcomes were obtained by using 1% sodium hydroxide (3 h) at room temperature, reducing acetyl and a number of uronic acid replacements that are blockers of sugar breakdown. This increased methane production by 34% (Shetty et al. 2017). When RS was alkali pretreated with 6% sodium hydroxide, outputs rose from 27% to 64% (He et al. 2008). With 6% sodium hydroxide and 5% sodium hydroxide as pretreatments, comparable biogas yields were obtained (He et al. 2008; Chen et al. 2010).

With 5% NaOH added after hydrothermal pretreatment, RS produced twice as much biomethane as untreated RS (Chandra et al. 2012). For lignocellulose, pretreatment with ammonia solution is beneficial. In contrast to carbohydrates, a solution of ammonia reacts with lignin in a very selective manner.

2.4.2.2 Acid Pretreatment

As lignin condenses and crystallizes during the acid pretreatment, hemicellulose is hydrolyzed into monosaccharides (Sousa et al. 2009; Salihu and Alam 2016). The method works well for materials with a high content of lignocellulose, because it degrades the lignin and hydrolytic microorganisms can adapt to an acidic environment (Salihu and Alam 2016). Enzyme activity significantly improved after acid pretreatment, which successfully eliminated hemicellulose (Chen et al. 2007). Because these acids are potent and aid in the hydrolysis of cellulose without the need of enzymes, they are frequently employed for strong acid pretreatment of lignocellulose with high concentrations of H_2SO_4 and HCl (Sun and Cheng 2002). Because these acids have corrosion-causing qualities, they should be recycled to cut down on pretreatment expenditures.

2.4.2.3 Hydrogen Peroxide (H_2O_2) Pretreatment

Song et al. (2013) conducted an experiment where RS was subjected to a pretreatment process at ambient temperature of 25 °C for 6.18 days. They used 2.68% hydrogen peroxide and a substrate/inoculum ratio of 1.08. This pretreatment resulted in the production of 288 mL/g VS (volatile solids) of methane. Notably, the hydrogen peroxides pretreatment led to significant reductions in cellulose (0.90% to 22.00%), hemicellulose (3.70% to 60.30%), and lignin (0.40% to 12.50%), respectively. These findings highlight the effectiveness of H_2O_2 pretreatment in modifying the composition of RS, leading to a higher methane yield during anaerobic digestion.

2.4.3 Biological Pretreatment

Since it uses less energy than physical pretreatment and is less costly in comparison to the chemical pretreatment, which is expensive since it calls for more expensive chemicals, biological pretreatment for AD has attracted interest. Due to its non-inhibitory nature and consideration for the environment, the biological approach appears to have the most promise. These pretreatment techniques shorten the time of AD to complete, improve the digestibility of the food, and boost gas liberation. This is because the lignocellulose of the straw are digested into simpler chemicals during AD, especially when utilizing microorganisms with a high capacity for lignocellulose degradation (Cianchetta et al. 2014). The reactivity of fungi like white, brown, and soft rot fungi is primarily related to biological therapy. Mustafa et al. (2016) reported that pretreatment of RS using *Trametes reesi* did not have a significant impact on lignin degradation. *Pleurotus ostreatus* and *Trametes reesi* pretreatments resulted in increased dry matter and degradation of cellulosic components. Additionally, both treatments demonstrated a higher degradation of hemicellulose and lignin compared to cellulose in RS. These findings suggest that *P. ostreatus* and *T. reesi* pretreatments can effectively modify the composition and breakdown of different components in RS. Comparing the pretreatment with *T. reesi* to raw RS, methane emission rose 9–78% at 65% and 75% content of moisture, but decreased 3–30% at 85%. There is significant potential when the incubation period and moisture content are chosen correctly.

2.4.4 Combined Pretreatment (Chemical + Microwave Irradiation)

Microwave irradiation has a high heating efficiency and can occasionally speed up reactions and shorten their duration, which could result in significant energy savings; it has been widely employed in chemistry. However, microwave processing might produce inhibitors like phenolic compounds and furfural as a side consequence (Hsu et al. 2010).

To prevent the production of these inhibitors, it is crucial to manage the pretreatment environment. Because of this, microwaves have not been utilized alone to pretreat lignocellulosic biomass, but rather they have typically been employed to help heat, acid, or alkaline pretreatment at less temperatures deprived of affecting the benefits of the pretreatment (Zhu et al. 2005). Alkali (NaOH) and a microwave and 1% NaOH were employed by Zhu et al. (2005). The outcome showed that RS lost 44.6% of its weight and that its composition consisted of cellulose at 69.2%, lignin at 4.9%, and hemicellulose at 10.2%. According to the results, RS prepared using a microwave had a higher rate of hydrolysis and a higher concentration of glucose in the hydrolysate than one made using an alkali alone. Hydrolysis of pretreated RS with enzymes was also examined. The elimination ratios of lignin were reported to be 46.1% and 51.54% using organic acid and a combination of microwave and chemical treatments (CH_3COOH and CH_3CH_2COOH), sugar yields were reported to be 71.41% and 80.08% under optimal conditions, which included a 25% acid concentration, a solid-liquid ratio of 1:15, a microwave intensity of 230 W, and an irradiation time of 5 min (Gong et al. 2010). This shows that RS's enzymatic hydrolysis sugar production can be improved by pretreating the organic acid with a microwave.

2.5 Codigestion of RS with Different Substrate

For an AD to be successful, ideal working conditions must be provided. It is generally recognized that certain nutrient deficiencies can result in poor performance or even the blockage of a process (Nasr et al. 2012). In light of this, codigestion presents itself as a practical and affordable means of resolving some of the challenges encountered throughout this process (Yang et al. 2015). Codigestion is the simple process of combining two or more substrates, which balances nutrients, increases diversity of microbes, boosts capacity of buffers, and dilutes inhibitors to improve the utilization of nutrients. As a result, codigestion may produce more methane than monodigestion of substrates (Nasr et al. 2012).

Due to its high concentration of cellulose and hemicellulose, lignocellulose biomass is regarded in this context as a fantastic candidate for AD. But, when used alone, its high C/N proportion and complex biomass composition obstruct the process of AD (Giuliano et al. 2013). However, waste from animals has a significant organic nitrogen content, which could lead to an ammonia accumulation and possibly obstruct microbial activity due to a low C/N ratio (Abouelenien et al. 2014; Li et al. 2014). Digestion of lignocellulose or manure of animals leads to less reliable performance and decreased methane generation because of the incorrect C/N proportion. By obtaining an adequate ratio of C/N during the codigestion of lignocellulosic biomass with adequate animal manure, it is possible to boost the AD process and produce more methane (Ogunwande et al. 2008; Ye et al. 2013).

Table 2.2 RS codigestion with different substrates to improve AD process

Substrate	Temp (°C)	Working volume (L)	Methane yield increase by (%)	Reference
Cow manure	37 ± 10	2.5	5.8	Li et al. (2015)
Pig manure	55	20	32	Riya et al. (2018)
Pig manure	35 ± 0.5	9	37.8	Shen et al. (2018)
Pig manure + kitchen waste	37	2	71.67 (than individual substrate)	Ye et al. (2013)
Urea ammoniated rice straw + food waste	35	1	8.83	Zhang et al. (2018)

Table 2.2 illustrates how the AD process is enhanced by RS codification with various substrates.

2.6 L-AD (Liquid AD) and SS-AD

L-AD systems initiate biogas production quickly but are prone to foaming issues, whereas SS-AD systems take longer to produce significant methane yields (Lianhua et al. 2010). In the context of SS-AD, system instability can occur due to two main factors: nutritional imbalance and ammonia accumulation. These issues can hinder the efficient functioning of the AD process. Additionally, low methane production in SS-AD can be caused by delayed mass movement, which affects the availability of substrates for microbial degradation, as well as the recalcitrance of lignocellulosic feedstock, which may resist decomposition by microbial enzymes. These factors, identified by Li et al. (2011), contribute to challenges in achieving optimal methane production in SS-AD systems. In the L-AD system, the primary metabolic mechanism for methanogenesis is the hydrogenotrophic route (Zhou and Wen 2019).

By intensifying mixing and designing an optimum reactor system, foaming in liquid anaerobic codigestion of straw and manure of pig can be reduced (Li et al. 2015). Compared to the typical liquid mesophilic AD process, thermophilic and solid-state AD processes have different chemical and biological processes (Jiunn-Jyi et al. 1997).

2.7 Plants for AD of RS at Pilot and Farm Scale

Knowing the fundamentals requires bench-scale study, but developing and operating a farm-scale plant using such microcosms is challenging (Mussoline et al. 2012). A farm-scale facility must also be monitored in order to gather comprehensive data and develop a deeper grasp of technology (Mussoline et al. 2012; Verma et al. 2023). It also aids in determining operationally useful details, including the duration of the startup stage and the energy, supply of co-substrates, and whether the system is long-term sustainably viable (Mussoline et al. 2014).

A study was done on the Pilot scale, in which grass from the roadside was combined with manure and RS in three separate digesters for AD (André et al. 2019). The biogas yield obtained by André et al. (2019) was comparable with yield stated by Mussoline et al. (2012). The yield from the three digesters in André et al.'s study was 226 L/kg VS (volatile solids). The BIORIST project successfully operated a pilot-scale facility using an efficient AD method to treat RS and manure, as documented by Larsen et al. (2019). Farm-scale biogas facilities are well-suited for utilizing RS as a feedstock to recover energy and stimulate agricultural development. Further research is necessary to enhance our understanding of the process. Specifically, studies on the long-term viability of farm-scale AD plants are required, addressing aspects such as startup and ongoing expenses, as well as the logistical considerations of handling RS and other co-substrates.

2.8 Conclusion

When treated anaerobically, lignocellulosic residues can represent substantial sources of renewable biogas. For RS digestion, AD is a dependable method. At mesophilic and thermophilic temperatures, RS can be effectively digested. By balancing nutrients, reducing toxic chemicals, adjusting content of moisture, increasing buffer capacity, and improving system constancy, the use of an appropriate inoculum and co-substrate in the AD of RS improves biogas/methane yield.

The biogas/methane yield is also improved by careful consideration given to the mixing, OLR, etc. Pretreatments and codigestion appear to be the best solutions among all those put up to address the issues lignocellulose wastes present in the process. Inhibition must be taken into account in relation to the operational conditions.

Among the various physical and chemical pretreatment methods, milling and alkaline pretreatment using sodium hydroxide (NaOH) are commonly employed and worth further investigation. These pretreatment methods enhance solubility, reduce complexity, and improve biogas production. To establish a farm-scale biogas plant for handling RS, it is crucial to design and define operational parameters based on pilot-scale plants and trials. These steps will provide valuable insights and ensure the successful implementation of AD for RS at a larger scale. Theoretically, it is feasible to install biogas plants at a farm-scale that combine RS with other acceptable substrates like PWW or other livestock waste. The nearby farm will receive energy and nutrient recovery from it. Therefore, it is important to assess the plant's economic viability.

References

Abouelenien F, Namba Y, Kosseva MR, Nishio N, Nakashimada Y (2014) Enhancement of methane production from co-digestion of chicken manure with agricultural wastes. Bioresour Technol 159:80–87

Abudi ZN, Hu Z, Xiao B, Abood AR, Rajaa N, Laghari M (2016) Effects of pretreatments on thickened waste activated sludge and rice straw co-digestion: experimental and modeling study. J Environ Manag 177:213–222

Alengebawy A, Ran Y, Ghimire N, Osman AI, Ai P (2023) Rice straw for energy and value-added products in China: a review. Environ Chem Lett:1–32. https://doi.org/10.1007/s10311-023-01612-3

André L, Zdanevitch I, Pineau C, Lencauchez J, Damiano A, Pauss A, Ribeiro T (2019) Dry anaerobic co-digestion of roadside grass and cattle manure at a 60 L batch pilot scale. Bioresour Technol 289:121737

Angelidaki I, Karakashev D, Batstone DJ, Plugge CM, Stams AJM (2011) Biomethanation and its potential. In: Rosenzweig AC, Ragsdale SW (eds) Methods in enzymology: methods in methane metabolism, vol 494. Elsevier, San Diego, pp 327–351

Anjum R, Grohmann E, Krakat N (2017) Anaerobic digestion of nitrogen rich poultry manure: impact of thermophilic biogas process on metal release and microbial resistances. Chemosphere 168:1637–1647

Appels L, Baeyens J, Degrève J, Dewil R (2008) Principles and potential of the anaerobic digestion of waste-activated sludge. Prog Energy Combust Sci 34(6):755–781

Ariunbaatar J, Panico A, Frunzo L, Esposito G, Lens PN, Pirozzi F (2014) Enhanced anaerobic digestion of food waste by thermal and ozonation pretreatment methods. J Environ Manag 146: 142–149

Arvanitoyannis IS, Tserkezou P (2008) Corn and rice waste: a comparative and critical presentation of methods and current and potential uses of treated waste. Int J Food Sci Technol 43(6): 958–988

Atelge MR, Krisa D, Kumar G, Eskicioglu C, Nguyen DD, Chang SW, Atabani AE, Al-Muhtaseb AH, Unalan S (2020) Biogas production from organic waste: recent progress and perspectives. Waste Biomass Valorization 11:1019–1040

Cai Y, Hua B, Gao L, Hu Y, Yuan X, Cui Z, Zhu W, Wang X (2017) Effects of adding trace elements on rice straw anaerobic mono-digestion: focus on changes in microbial communities using high-throughput sequencing. Bioresour Technol 239:454–463

Campanaro S, Treu L, Kougias PG, De Francisci D, Valle G, Angelidaki I (2016) Metagenomic analysis and functional characterization of the biogas microbiome using high throughput shotgun sequencing and a novel binning strategy. Biotechnol Biofuels 9:1–17

Chandra R, Takeuchi H, Hasegawa T (2012) Hydrothermal pretreatment of rice straw biomass: a potential and promising method for enhanced methane production. Appl Energy 94:129–140

Chen Y, Sharma-Shivappa RR, Chen C (2007) Ensiling agricultural residues for bioethanol production. Appl Biochem Biotechnol 143:80–92

Chen GY, Zheng Z, Luo Y, Zou XX, Fang CX (2010) Effect of alkaline treatment on anaerobic digestion of rice straw. Huan Jing Ke Xue 31(9):2208–2213

Choong YY, Norli I, Abdullah AZ, Yhaya MF (2016) Impacts of trace element supplementation on the performance of anaerobic digestion process: a critical review. Bioresour Technol 209: 369–379

Cianchetta S, Di Maggio B, Burzi PL, Galletti S (2014) Evaluation of selected white-rot fungal isolates for improving the sugar yield from wheat straw. Appl Biochem Biotechnol 173: 609–623

Clark RH, Speece RE (1971) The pH tolerance of anaerobic digestion. Adv Water Pollut Res 1: 1–13

Forster-Carneiro T, Pérez M, Romero LI (2008) Influence of total solid and inoculum contents on performance of anaerobic reactors treating food waste. Bioresour Technol 99(15):6994–7002

Gerardi MH (2003) The microbiology of anaerobic digesters. John Wiley & Sons, Hoboken, NJ
Gharpuray MM, Lee YH, Fan LT (1983) Structural modification of lignocellulosics by pretreatments to enhance enzymatic hydrolysis. Biotechnol Bioeng 25(1):157–172
Giuliano A, Bolzonella D, Pavan P, Cavinato C, Cecchi F (2013) Co-digestion of livestock effluents, energy crops and agro-waste: feeding and process optimization in mesophilic and thermophilic conditions. Bioresour Technol 128:612–618
Gollakota KG, Meher KK (1988) Effect of particle size, temperature, loading rate and stirring on biogas production from castor cake (oil expelled). Biol Wastes 24(4):243–249
Gong G, Liu D, Huang Y (2010) Microwave-assisted organic acid pretreatment for enzymatic hydrolysis of rice straw. Biosyst Eng 107(2):67–73
Guillaume S, Lendormi T (2015) Anaerobic co-digestion of dairy cattle slurry and agro-industrial-fats: effect of fat ratio on the digester efficiency. Can J Chem Eng 93(2):304–308
Hagos K, Zong J, Li D, Liu C, Lu X (2017) Anaerobic co-digestion process for biogas production: progress, challenges and perspectives. Renew Sust Energ Rev 76:1485–1496
Haider MR, Yousaf S, Malik RN, Visvanathan C (2015) Effect of mixing ratio of food waste and rice husk co-digestion and substrate to inoculum ratio on biogas production. Bioresour Technol 190:451–457
Hansen KH, Angelidaki I, Ahring BK (1998) Anaerobic digestion of swine manure: inhibition by ammonia. Water Res 32(1):5–12
Harmsen PF, Huijgen W, Bermudez L, Bakker R (2010) Literature review of physical and chemical pretreatment processes for lignocellulosic biomass (no. 1184). Wageningen UR-Food & Biobased Research
Hashimoto AG (1986) Ammonia inhibition of methanogenesis from cattle wastes. Agric Wastes 17(4):241–261
He Y, Pang Y, Liu Y, Li X, Wang K (2008) Physicochemical characterization of rice straw pretreated with sodium hydroxide in the solid state for enhancing biogas production. Energy Fuel 22(4):2775–2781
Hendriks ATWM, Zeeman G (2009) Pretreatments to enhance the digestibility of lignocellulosic biomass. Bioresour Technol 100(1):10–18
Hsu TC, Guo GL, Chen WH, Hwang WS (2010) Effect of dilute acid pretreatment of rice straw on structural properties and enzymatic hydrolysis. Bioresour Technol 101(13):4907–4913
Hu F, Ragauskas A (2012) Pretreatment and lignocellulosic chemistry. Bioenergy Res 5:1043–1066
Jiunn-Jyi L, Yu-You L, Noike T (1997) Influences of pH and moisture content on the methane production in high-solids sludge digestion. Water Res 31(6):1518–1524
Kabir MM, Forgács G, Sárvári Horváth I (2015) Biogas from lignocellulosic materials. In: Karimi K (ed) Lignocellulose-based bioproducts. Springer, Cham, pp 207–251
Larsen O, Rotter VS, Fechter L, Spahr M (2019) BioRist-Entwicklung und Integration eines innovativen Verfahrens zur Biogasherstellung aus Reisstroh in regionale Wertschöpfungsketten im ländlichen Raum in Südostasien unter Berücksichtigung nachhaltiger Entwicklung und Klimaschutz–Beispiel Vietnam
Li Y, Park SY, Zhu J (2011) Solid-state anaerobic digestion for methane production from organic waste. Renew Sust Energ Rev 15(1):821–826
Li Y, Zhang R, He Y, Zhang C, Liu X, Chen C, Liu G (2014) Anaerobic co-digestion of chicken manure and corn Stover in batch and continuously stirred tank reactor (CSTR). Bioresour Technol 156:342–347
Li D, Liu S, Mi L, Li Z, Yuan Y, Yan Z, Liu X (2015) Effects of feedstock ratio and organic loading rate on the anaerobic mesophilic co-digestion of rice straw and cow manure. Bioresour Technol 189:319–326
Lianhua L, Dong L, Yongming S, Longlong M, Zhenhong Y, Xiaoying K (2010) Effect of temperature and solid concentration on anaerobic digestion of rice straw in South China. Int J Hydrog Energy 35(13):7261–7266

Mancini G, Papirio S, Riccardelli G, Lens PN, Esposito G (2018) Trace elements dosing and alkaline pretreatment in the anaerobic digestion of rice straw. Bioresour Technol 247:897–903

McCarty PL (1964) Anaerobic waste treatment fundamentals. Public Works 95(9):107–112

Monnet F (2003) An introduction to anaerobic digestion of organic wastes. Remade Scotland 379: 1–48

Mshandete A, Björnsson L, Kivaisi AK, Rubindamayugi MS, Mattiasson B (2006) Effect of particle size on biogas yield from sisal fibre waste. Renew Energy 31(14):2385–2392

Mussoline W, Esposito G, Lens P, Garuti G, Giordano A (2012) Design considerations for a farm-scale biogas plant based on pilot-scale anaerobic digesters loaded with rice straw and piggery wastewater. Biomass Bioenergy 46:469–478

Mussoline W, Esposito G, Lens P, Garuti G, Giordano A (2014) Electrical energy production and operational strategies from a farm-scale anaerobic batch reactor loaded with rice straw and piggery wastewater. Renew Energy 62:399–406

Mustafa AM, Poulsen TG, Sheng K (2016) Fungal pretreatment of rice straw with Pleurotus ostreatus and trichoderma reesei to enhance methane production under solid-state anaerobic digestion. Appl Energy 180:661–671

Nagao N, Tajima N, Kawai M, Niwa C, Kurosawa N, Matsuyama T, Yusoff FM, Toda T (2012) Maximum organic loading rate for the single-stage wet anaerobic digestion of food waste. Bioresour Technol 118:210–218

Nair A (2013) Effect of leachate blending on anaerobic digestion of organic fraction of municipal solid waste. University of Ottawa, Ottawa

Nasr N, Elbeshbishy E, Hafez H, Nakhla G, El Naggar MH (2012) Comparative assessment of single-stage and two-stage anaerobic digestion for the treatment of thin stillage. Bioresour Technol 111:122–126

Ngan NVC, Chan FMS, Nam TS, Van Thao H, Maguyon-Detras MC, Hung DV, Cuong DM, Van Hung N (2020) Anaerobic digestion of rice straw for biogas production. In: Gummert M, Hung N, Chivenge P, Douthwaite B (eds) Sustainable rice straw management. Springer, Cham, pp 65–92

Ogunwande GA, Osunade JA, Adekalu KO, Ogunjimi LAO (2008) Nitrogen loss in chicken litter compost as affected by carbon to nitrogen ratio and turning frequency. Bioresour Technol 99(16):7495–7503

Pore SD, Shetty D, Arora P, Maheshwari S, Dhakephalkar PK (2016) Metagenome changes in the biogas producing community during anaerobic digestion of rice straw. Bioresour Technol 213: 50–53

Rajagopal R, Massé DI, Singh G (2013) A critical review on inhibition of anaerobic digestion process by excess ammonia. Bioresour Technol 143:632–641

Riya S, Suzuki K, Meng L, Zhou S, Terada A, Hosomi M (2018) The influence of the total solid content on the stability of drythermophilic anaerobic digestion of rice straw and pig manure. Waste Manag 76:350–356

Salihu A, Alam MZ (2016) Pretreatment methods of organic wastes for biogas production. J Appl Sci 16(3):124–137

Sawatdeenarunat C, Surendra KC, Takara D, Oechsner H, Khanal SK (2015) Anaerobic digestion of lignocellulosic biomass: challenges and opportunities. Bioresour Technol 178:178–186

Schmidt JE, Ahring BK (1993) Effects of hydrogen and formate on the degradation of propionate and butyrate in thermophilic granules from an upflow anaerobic sludge blanket reactor. Appl Environ Microbiol 59(8):2546–2551

Shen F, Li H, Wu X, Wang Y, Zhang Q (2018) Effect of organic loading rate on anaerobic co-digestion of rice straw and pig manure with or without biological pretreatment. Bioresour Technol 250:155–162

Shetty DJ, Kshirsagar P, Tapadia-Maheshwari S, Lanjekar V, Singh SK, Dhakephalkar PK (2017) Alkali pretreatment at ambient temperature: a promising method to enhance biomethanation of rice straw. Bioresour Technol 226:80–88

Song ZL, Yag GH, Feng YZ, Ren GX, Han XH (2013) Pretreatment of rice straw by hydrogen peroxide for enhanced methane yield. J Integr Agric 12(7):1258–1266
Soto M, Méndez R, Lema JM (1993) Methanogenic and non-methanogenic activity tests. Theoretical basis and experimental set up. Water Res 27(8):1361–1376
Sousa LD, Chundawat SP, Balan V, Dale BE (2009) 'Cradle-to-grave' assessment of existing lignocellulose pretreatment technologies. Curr Opin Biotechnol 20(3):339–347
Sreekrishnan TR, Kohli S, Rana V (2004) Enhancement of biogas production from solid substrates using different techniques—a review. Bioresour Technol 95(1):1–10
Sträuber H, Schröder M, Kleinsteuber S (2012) Metabolic and microbial community dynamics during the hydrolytic and acidogenic fermentation in a leach-bed process. Energy Sustain Soc 2(1):1–10
Sun Y, Cheng J (2002) Hydrolysis of lignocellulosic materials for ethanol production: a review. Bioresour Technol 83(1):1–11
Taherzadeh MJ, Karimi K (2008) Pretreatment of lignocellulosic wastes to improve ethanol and biogas production: a review. Int J Mol Sci 9(9):1621–1651
Takdastan A, Movahedian H, Jafarzadeh N, Bina B (2005) The efficiency of anaerobic digesters on microbial quality of sludge in Isfahan and Shahinshahr water waste treatment plant. J Environ Health Sci Eng 2(1):56–59
Treu L, Campanaro S, Kougias PG, Zhu X, Angelidaki I (2016) Untangling the effect of fatty acid addition at species level revealed different transcriptional responses of the biogas microbial community members. Environ Sci Technol 50(11):6079–6090
Tsapekos P (2017) Enhancing biogas production from recalcitrant lignocellulosic residue. Unpublished Ph. D thesis, Technical University of Denmark, Kongens Lyngby, Denmark
Verma D, Sahni T, Kumar S, Sarao LK (2023) Synthesis and application of nanoengineered cellulosic biomass in biohydrogen production. In: Srivastava M, Mishra PK (eds) Nanobioenergy: application and sustainability assessment. Clean energy production technologies. Springer, Singapore, pp 135–167. https://doi.org/10.1007/978-981-19-6234-9_6
Ward AJ, Hobbs PJ, Holliman PJ, Jones DL (2008) Optimisation of the anaerobic digestion of agricultural resources. Bioresour Technol 99(17):7928–7940
Yang L, Xu F, Ge X, Li Y (2015) Challenges and strategies for solid-state anaerobic digestion of lignocellulosic biomass. Renew Sust Energ Rev 44:824–834
Ye J, Li D, Sun Y, Wang G, Yuan Z, Zhen F, Wang Y (2013) Improved biogas production from rice straw by co-digestion with kitchen waste and pig manure. Waste Manag 33(12):2653–2658
Yerkes DW, Boonyakitsombut S, Speece RE (1997) Antagonism of sodium toxicity by the compatible solute betaine in anaerobic methanogenic systems. Water Sci Technol 36(6-7): 15–24
Zealand AM, Roskilly AP, Graham DW (2017) Effect of feeding frequency and organic loading rate on biomethane production in the anaerobic digestion of rice straw. Appl Energy 207:156–165
Zhang R, Zhang Z (1999) Biogasification of rice straw with an anaerobic-phased solids digester system. Bioresour Technol 68(3):235–245
Zhang X, Cheng S, Wang X, Huang X, Logan BE (2009) Separator characteristics for increasing performance of microbial fuel cells. Environ Sci Technol 43(21):8456–8461
Zhang H, Luo L, Li W, Wang X, Sun Y, Sun Y, Gong W (2018) Optimization of mixing ratio of ammoniated rice straw and food waste co-digestion and impact of trace element supplementation on biogas production. J Mater Cycles Waste Manage 20:745–753
Zhao R, Zhang Z, Zhang R, Li M, Lei Z, Utsumi M, Sugiura N (2010) Methane production from rice straw pretreated by a mixture of acetic–propionic acid. Bioresour Technol 101(3):990–994
Zhou H, Wen Z (2019) Solid-state anaerobic digestion for waste management and biogas production. In: Steudler S, Werner A, Cheng J (eds) Solid sate fermentation: research and industrial applications. Springer, Cham, pp 147–168
Zhu S, Wu Y, Yu Z, Liao J, Zhang Y (2005) Pretreatment by microwave/alkali of rice straw and its enzymic hydrolysis. Process Biochem 40(9):3082–3086

Luo L, Gong W, Qin L, Ma Y, Ju W, Wang H (2018) Influence of liquid-and solid-state coupling anaerobic digestion process on methane production of cow manure and rice straw. Journal of Material Cycles and Waste Management 20:1804–1812

Meng L, Xie L, Kinh CT, Suenaga T, Hori T, Riya S, Terada A, Hosomi M (2018) Influence of feedstock-toinoculum ratio on performance and microbial community succession during solid-state thermophilic anaerobic codigestion of pig urine and rice straw. Bioresource Technology 252:127–133

Chapter 3
A Wealthy Green Approach: Conversion of Paddy Straw Waste into Biorefinery Products

Neha Sharma, Manmeet Kaur, Aneeshpal Singh, Sandeep Kaur, Diksha Singla, and Ravish Choudhary

Abstract The need to produce more food has increased the amount of agro-industrial waste produced. Although this was generally recognized as a major problem in the past, it is now believed that the development of sustainable industrial processes depends heavily on this waste. Microbial transformations based on biotechnology provide effective, affordable, and sustainable approaches for producing goods with additional value. The production of nutrient-rich raw materials and biorefineries could use agricultural waste as a sustainable resource. The prospective use of agricultural residues for the synthesis of by-products such as levulinic acid, xylose, lactic acid, and sophorolipids, adsorbents, in the construction sector and for biofuels using numerous bioconversion methods is the main focus of this chapter. It is clear that the use of crop residues that have withstood bioconversion as fuel is an amazing area of research with enormous potential for massive-scale agro-industrial production to effectively meet growing energy needs.

Keywords Agriculture waste · Biorefinery products · Ecofriendly · Paddy straw · Sustainable development

N. Sharma · M. Kaur (✉)
Department of Microbiology, Punjab Agricultural University, Ludhiana, Punjab, India
e-mail: manmeet1-mb@pau.edu

A. Singh
Department of Food Science and Technology, Punjab Agricultural University, Ludhiana, Punjab, India

S. Kaur
Department of Entomology, Punjab Agricultural University, Ludhiana, Punjab, India

D. Singla
Department of Biochemistry, Punjab Agricultural University, Ludhiana, Punjab, India

R. Choudhary
Division of Seed Science and Technology, Indian Agricultural Research Institute, Delhi, India

N. Srivastava et al. (eds.), *Paddy Straw Waste for Biorefinery Applications*, Clean Energy Production Technologies, https://doi.org/10.1007/978-981-99-8224-0_3

3.1 Introduction

Ninety percent of the rice consumed worldwide is produced exclusively on the Asian continent. Over 700 million tons of rice is consumed worldwide each year, making it the second most popular food. The amount of rice straw produced per ton of rice grains on agricultural land is said to be between 1.36 and 1.51 t. For every ton of rice processed in the industry, rice husks, broken rice, and rice bran are produced. Agriculture and the food industry produce large amounts of waste that can be utilized as raw substrate for the development and production of biorefinery substances, offering numerous potential production opportunities that are environmentally beneficial (Ng et al. 2020). Because improper disposal could pose both a hazard to human health and to the environment, the production of solid rice waste has become a serious problem. Rice straw is one of these pollutants that are often burned in the fields because it is relatively easy to remove, leading to environmental pollution, dermatological problems, and respiratory diseases. In addition, burning rice straw outdoors releases a significant amount of greenhouse gases into the air, including carbon monoxide, nitrous oxide, sulfur oxide, methane, and particulates (Singh and Patel 2022). The solid-state fermentation process can convert rice straw, which is inherently lignocellulose, into a number of other products (Gunjan et al. 2023). In addition, cellulosic polymers produced from agricultural and industrial wastes can be used to produce new bioproducts such as hydrogels, bio-nanocomposites, bioaerogels, biofilms, and some biomaterials which are majorly utilized in tissue engineering (Soorbaghi et al. 2019). As shown in Fig. 3.1, the by-products of the rice straw waste are utilized in the various biorefining processing stages. In this case, high quality and yield of the final extract produced by eco-friendly extraction methods have become a fantastic alternative for recovering high-value products from agro-industrial waste (Soquetta et al. 2018). Advanced technologies have been applied for the production of biorefinery products from the paddy straw including ultrasound-assisted extraction (UAE), supercritical fluid extraction (SFE), and microwave-assisted extraction (MAE) (Angoy et al. 2020). In addition, compared to previous technologies, these are less energy intensive, require fewer organic solvents, and work faster (Sik et al. 2020). The primary goal of upcoming chapter is to provide a brief description of the biorefinery products made from rice straw and the synthesis of chemical compounds from this biomass with significant potential applications.

3.2 Microbial Bioprocessing of Paddy Straw

Microbial bioprocessing technology holds a unique place in waste management plans around the world as it promotes sustainable growth and offers an environmentally friendly alternative. The two fermentation-based methods used in this process are submerged and solid-state fermentation. As shown in Fig. 3.2, microorganisms

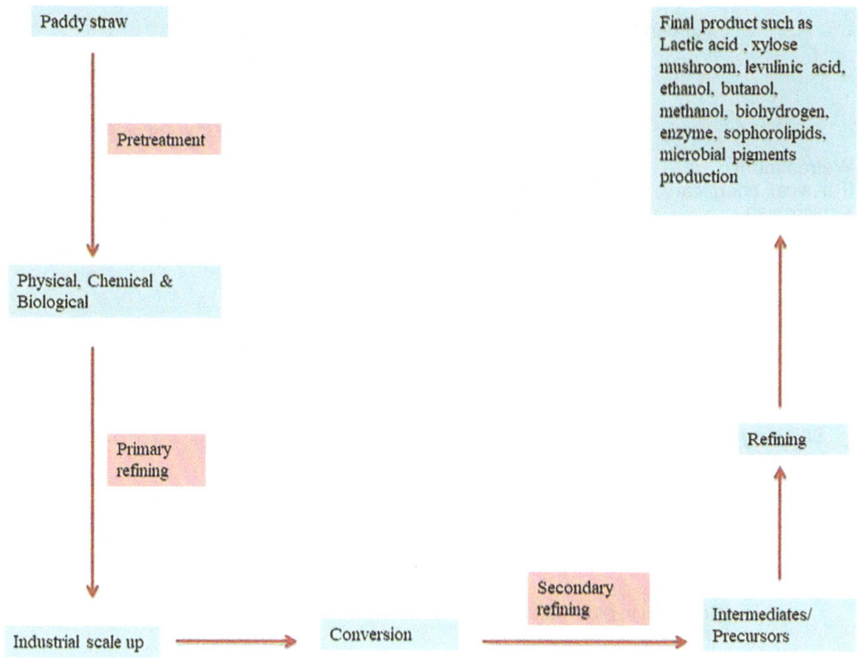

Fig. 3.1 Synthesis of biorefinery products from paddy straw

grow and produce biorefinery products in both processes by using agro-industrial residues as a food source (Sodhi et al. 2022).

3.2.1 Submerged Fermentation (SmF)

A technique for microbial growth in suspension cultures, called submerged fermentation, allows by-products to be released into the surrounding broth. Large volumes can now be managed more successfully, and process variables such as dissolved oxygen, pH, pressure, and temperature can be controlled in a better way (Pinotti et al. 2020).

3.2.2 Solid-State Fermentation (SSF)

Both industry and academia devote a great deal of attention to solid-state fermentation (SSF). Because SSF consumes almost no water, it closely resembles the microorganism's natural environment. Comparing this method to SmF, Ring et al.

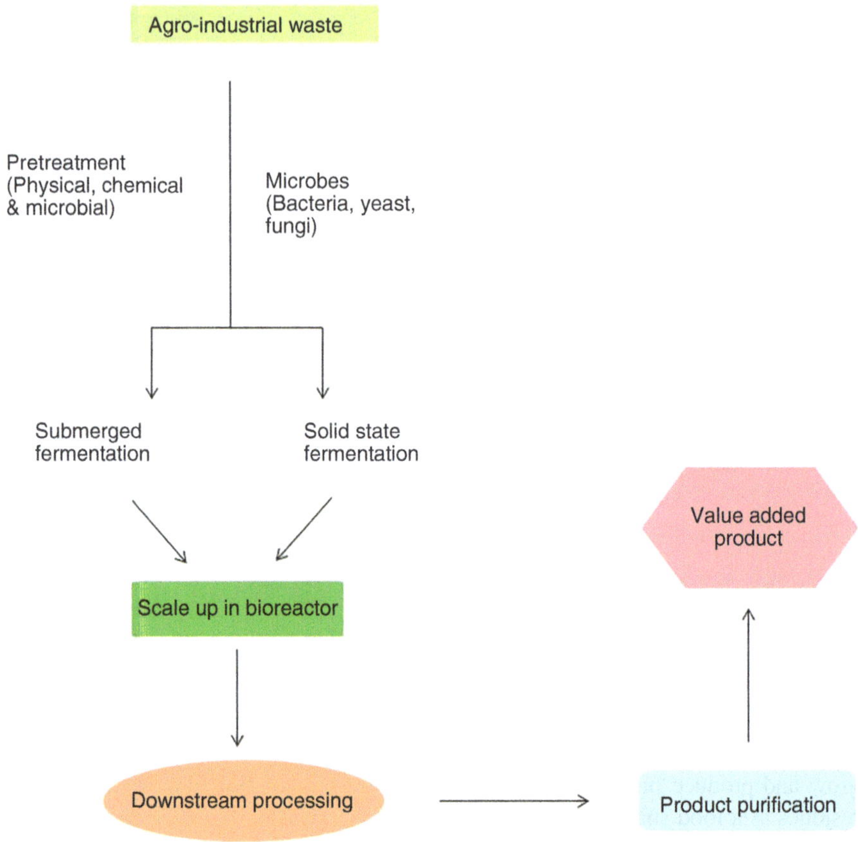

Fig. 3.2 Diagrammatic representation of microbial processing of agro-industrial waste

(2018) found that it provides better solid support for optimal microbial development, ability to use complicated substrates in their natural state, and low relative humidity.

3.3 Advanced Extraction Technologies Applied to Paddy Straw

The extraction and processing of extremely valuable materials from industrial and agricultural waste is a new breakthrough in environmentally friendly processes. The current trend is that novel extraction technologies are gradually replacing time-consuming, expensive, and often organic solvent-intensive conventional processes. Because they are more widely applicable, more selective in terms of the commodities extracted, and more efficient overall, these environmentally friendly extraction techniques can support the sustainability of the industry (Belwal et al. 2020).

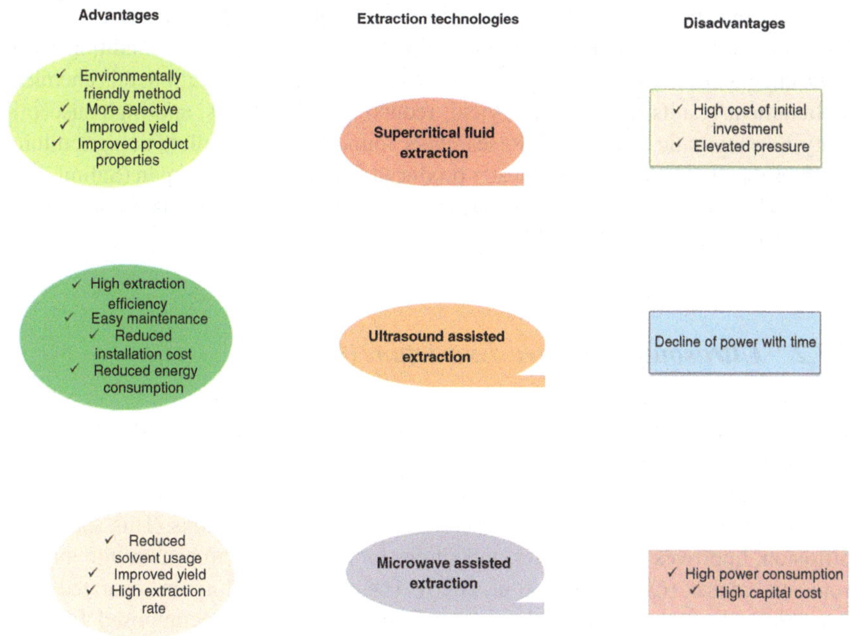

Fig. 3.3 Advantages and disadvantages of greener extraction technologies

Figure 3.3 displays the benefits and drawbacks of several ecologically friendly extraction methods. Numerous studies on ecologically friendly extraction techniques have recently been carried out all around the world. Supercritical fluid extraction (SFE), ultrasonic-assisted extraction (UAE), and microwave-assisted extraction (MAE) stand out among these cutting-edge methods.

3.3.1 *Supercritical Fluid Extraction (SFE)*

SFE refers to alterations in crucial variables such as pressure and temperature which can transform a gaseous form into a supercritical medium with barely perceptible liquid and gaseous phases. The main transport route in this mass transfer process is the frequent turbulence in the saturated solvent phase (Soquetta et al. 2018). Since it allows the extraction of various substances enriched with antioxidant properties and also have numerous industrial applications, therefore supercritical fluid extraction is considered an interesting way to utilize lignocellulosic wastes (Lizcano et al. 2019). Biodiesel production is being increased by supercritical fluid technology. Researchers used supercritical methanol and propane as co-solvents to produce tons of biodiesel annually using this method (Ortiz and de Santa-Ana 2017). Pretreating rice stalks with subcritical carbon dioxide ethanol reduced the amount of lignin from 23.01% to 21.42% (Yang et al. 2018). An oil yield of 17.19% was achieved by supercritical CO_2 extraction over 120 min (Moreira et al. 2023). Supercritical fluids can pretreat lignocellulosic residues under mild conditions with some chemicals including catalysts, solvents, and reagents. In addition, this chemical pretreatment results in high sugar yield, reduction in inhibitors, and high susceptibility to enzymatic hydrolysis. Researchers found that lignocellulosic agricultural residues gasified in supercritical water produced 5.56 mmol/g hydrogen (Salimi et al. 2016). The oil extracted *G. glabra* using this extraction method was found to have maximum yield around 79% (Durak 2014).

3.3.2 Ultrasound-Assisted Extraction (UAE)

The UAE seems to be one of the least expensive unconventional extraction techniques. Therefore, extensive research work has been carried out to make it practical on an industrial scale. When cell walls break down, target components can be released. This is achieved through the use of ultrasonic vibrations. This speeds up diffusion, improves mass transfer, and allows for greater solvent penetration into the sonicated matrix (Chakraborty et al. 2020). The cellulose and lignin separated from rice straw showed improved thermal resistance; substrates decomposed faster at higher temperatures. The UAE alkaline method has been proposed for the extraction of lignin and cellulose from paddy straw (Vu et al. 2017). The cellulosic fiber yield of the rice straw increased to 37% when heated by the ultrasonic reflux method. Sonication has a direct impact on the crude enzyme extraction process from the medium and hence showed that activity of industrial enzymes such as pectinase (1.20-fold), carboxymethylcellulase (1.48-fold), and xylanase (1.30-fold) increased (Li et al. 2015). Yang and Fang (2014) observed that ultrasonic assisted extraction method induced the enzymatic hydrolysis of acidic treated rice straw and resulted in conversion of total sugars (96.21%). Gallic acid (342 mg) was extracted from rice husks using an ultrasonic extraction method (Freitas et al. 2022).

3.3.3 Microwave-Assisted Extraction (MAE)

Microwave radiation has a frequency range of 30 MHz to 300 GHz (Chen et al. 2017). MAE offers a number of advantages over traditional extraction techniques, such as selectivity of extraction, easy maintenance, lower installation costs, less time spent on process setup, shorter extraction times, and lower power consumption. Rice straw treated with MAE showed a very high crystallinity index (54.6%) by using X-ray diffraction (Singh et al. 2014). Under optimal microwave intensity conditions, maximum efficiencies improved for cellulose (31.2%), hemicellulose (43.3%), and total saccharification (30.3%) (Ma et al. 2009). During the enzymatic hydrolysis of Chakhao Poireiton black rice, microwave treatment increased the release of glucose by 25–50% (Moirangthem et al. 2021). The laccase enzyme isolated from *Trametes hirsuta* MTCC-1171 showed yield (53.9%) during the treatment of enzyme by microwave-assisted three-phase partitioning (Patil and Yadav 2018).

3.4 Utilization of Paddy Straw Biomass: Bioconversion to Biorefinery Products

3.4.1 Levulinic Acid (LA) Production

Among the 12 molecular substrates that create value-added products is levulinic acid (Kim and Han 2021). Hexoses such as fructose, galactose, and glucose are hydrolyzed to produce LA. The high production interest is due to the chemical composition of LA, which includes both an acidic and a ketone carbonyl group. This blueprint can be utilized in healthcare and agro-food industry. Important compounds such as 2-methyltetrahydrofuran, acrylic acid, and 1, 4-pentanodiol can be industrially produced using levulinic acid (Emelyanenko et al. 2018). Cañon et al. (2022) reported that the overall output of ethyl levulinate per kilogram of rice straw was 0.115 kg and the total energy input was 133.2 kWh. The aqueous Soxhlet extraction of pretreated rice hulls produced the maximum yield (59.4%) of levulinic acid (Bevilaqua et al. 2013). The yield of LA from ultrafine ground steam-exploded rice straw was 70% (Yaaini et al. 2012). Untreated rice straw yielded 49.8% LA with an enhanced catalytic thermal liquefaction process (Ukarde and Pawar 2022).

3.4.2 Lactic Acid Production

Lactic acid and its derivatives are numerously utilized in multiple industries, including agri-food sector, medical, healthcare products, and cosmetics, followed by dermatological products. Polylactic acid is a biocompatible and biodegradable plastic that has recently gained importance as an unprocessed product (Eş et al. 2018). The majority of the cost of lactic acid production comes from the composition of the medium, although the economics of such process is still difficult to understand. To reduce the overall price of lactic acid production, numerous methods have been used to hydrolyze cellulosic carbohydrates (Ahmad et al. 2020). Bioconversion of agro-industrial residues products requires the hydrolysis of pretreated lignocellulosic materials into fermentable sugars (Kaur et al. 2023). Addition of polyoxyethylene sorbitan monooleate (PSM) resulted in maximum acid (L-lactic) production (Yao et al. 2007). From raw rice straw, *Bacillus coagulans* can extract up to 64.03 g/L and 297.02 g/kg of lactic acid (Chen et al. 2019). According to Qi and Yao (2007), rice straw could produce a maximum of 3.5 g of L-lactic acid.

3.4.3 Xylose Production

Since lignocellulosic wastes are accessible, affordable, and renewable sources of sugar, there is growing interest in biotechnological approaches to utilizing these wastes. A technically and economically effective bioprocess for xylitol synthesis is often developed using hemicellulose hydrolysates obtained from waste materials. Seventy-seven percent of the xylose was recovered from the hydrolysis of paddy straw (Roberto et al. 2003). The rice starch hydrolyzate treated with charcoal showed highest xylitol (0.53 g/g) (Baek and Kwon 2007). After chemical pretreatment of rice straw, 14.04 g of crystalline xylose was formed (Krishania et al. 2018). *Candida tropicalis* GS18 produced xylitol titers from paddy straw ranging between 25 and 35 g/L (Kaur et al. 2022b). The fermentation of paddy straw hydrolyzate with *C. tropicalis* resulted in production of 0.6 g/g of xylose-xylitol (Singh et al. 2021b). During consolidated bioprocessing, nanofiltration of the rice straw hydrolyzate allowed the elimination of inhibitors and allowed maximum xylitol synthesis (38.1 g/L) (Guirimand et al. 2016). Lignocellulosic residues has the ability to yield maximum xylitol ranging between 13 and 97 g/L (Azizah 2019).

3.4.4 Adsorbents for Contamination

Because it is a readily available and affordable substrate, lignocellulosic biomass such as paddy straw play an integral role and act as an adsorbent to eliminate heavy metals from wastewater. The highest results for raw rice straw during batch fermentation were 20.9% and 8.4%, respectively. Adsorbents such as raw, alkaline, and charcoal rice straw showed stronger adsorption of cadmium ions than adsorption of zinc ions. According to Singh et al. (2016), rice straw powder has maximum adsorption capacities for methylene blue, phosphorus, and nickel of 22.01, 4.2, and 4.01 mg/g, respectively. Using a silica-based adsorbent for rice straw, adsorption values were calculated for nitrate levels of approximately 15.04 mg/g and for groundwater samples of 8.54 mg/g (Robles-Jimarez et al. 2022). Both rice and straw husks absorbed fluorine in a range of 8–15 mg/g. According to Collivignarelli et al. (2022), rice husks have an arsenic adsorption capacity of over 19 mg/g. At appropriate dosage of rice straw, the highest chromium adsorption capacity was 15.80 mg/g (Wu et al. 2016). Rice straw charcoal was able to absorb the largest amount of tetracycline from an aqueous solution (Fan et al. 2018). Congo red and methylene blue are degraded by rice straw nanocomposites with adsorption capacities of 399 and 410 mg/g (Sangon et al. 2021).

3.4.5 Alternative Materials in Construction Industry

Due to population growth and demand for better building environments, the construction industry is expected to generate more greenhouse gas emissions. Carbon emissions can no longer be offset by the Earth's carbon cycle. In order to stop the emission of carbon dioxide from industrial processes such as the construction sector, it is therefore imperative to use technologies that can absorb and store carbon dioxide. This is required to drastically reduce the carbon footprint of the cement processing sectors and bring about positive environmental changes. Paddy straw has the potential to be a very effective carbon dioxide sorption material in cement-based applications (Akinyemi et al. 2020). High-performance and environmentally friendly construction walls can be made using bales of rice straw. Different concrete strengths can be increased by utilizing rice straw ash (7%) and microsilica (24%), respectively (Pandey and Kumar 2020). According to Jittin et al. (2020), the addition of rice hull ash increases perceived porosity, which is critical to reducing brick density. This allows the construction of infill walls from such bricks, thereby reducing the overall mass of the entire structure (De Silva and Perera 2018).

3.4.6 Applications in Renewable Energy

Rice straw is an interesting resource for bioenergy production due to its accessibility and abundance. It is one of the important crops that provide significant amounts of silica-enriched biomass (Rathour et al. 2023). One of the major challenges in converting rice straw into a biofuel that promotes green bioenergy development is the chemical structure of the feedstock (Kaur et al. 2023). The ethanol yield and concentration using the two-stage treated rice straw were reported to be 86.41% and 41.03 g/L, respectively, after 72 h of fermentation (Zhu et al. 2015). Nandal et al. (2020) found that *P. stipitis* NCIM3497 and *P. stipitis* NCIM3498 had the highest fermentation efficiencies for rice straw hydrolyzate within 24 h (57.30% and 53.02%, respectively). At an ethanol concentration (45% v/v), the straw produced the hydrogen emission up to 19.70 mL/g (Asadi et al. 2017). Using co-substrates of 40% hydrolyzed rice straw and 60% fermentation culture supernatant, acetone-butanol-ethanol fermentation resulted in hydrogen and butanol yields of 5897 mL/L and 14.1 g/L (Li et al. 2018). The synthesis of butanol increased significantly to 7.1 g/L after cellulase was introduced into delignified rice straw (Kiyoshi et al. 2015). The rice straw treated with sodium hydroxide yielded 28.05 g/L reducing sugars. *Saccharomyces tanninophilus* produced ethanol with maximum yield of 84.1% using the different hydrolysis-fermentation techniques (Jin et al. 2020).

3.4.7 Mushroom Production

One of the most cost-effective ways to recycle organic materials is to grow special mushrooms on lignocellulosic waste as shown in Fig. 3.4 (Kaur et al. 2022a). The most popular substrate for the mushroom industry is rice straw. According to Moonmoon et al. (2010), *Pleurotus eryngii* on rice straw had a biological yield and efficiency of 68.7%. Rice straw has remarkable potential for growing *Ganoderma lucidum*, evidenced by the fact that it could be grown as a primary substrate (Veena and Pandey 2011). Oyster mushrooms grow faster on rice straw, with less mycelial density on the surface, faster production of primordia, lower yield, and faster mycelial growth rate. With rice straw as a substrate, *Volvariella volvacea* (Bull.) achieved the best biological efficiency (7.93%) (Zikriyani et al. 2018).

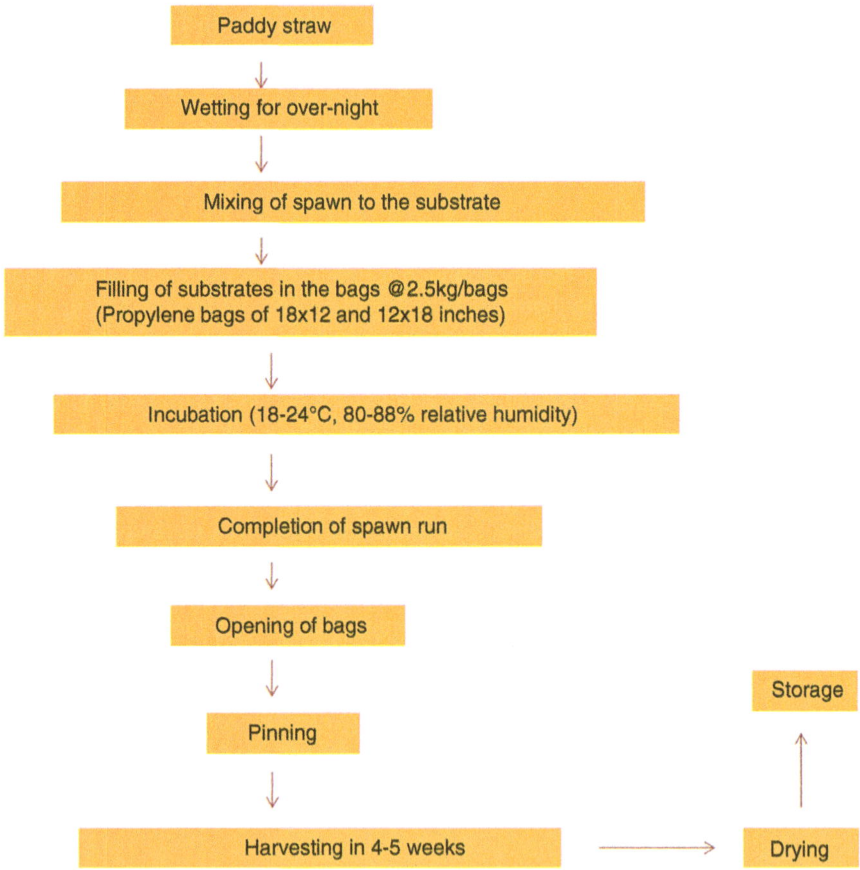

Fig. 3.4 Schematic representation of mushroom production using lignocellulosic waste

3.4.8 Production of Sophorolipids

Oleic acid, carbohydrates, and other unconventional substrates, including agro-industrial residues, can be used for biosurfactant production (Celligoi et al. 2020). The cost of raw materials accounts for 10–30% of the total price for the production of sophorolipids (Zaman and Hamid 2022). Therefore, the use of low-cost substrates can be a choice that increases the accessibility of the production process while reducing its negative impact on the environment (Satpute et al. 2017). Because of their exceptional ability to promote emulsification, solubilization, and lowering interfacial tension, sophorolipids are bioactive substances that have recently received much attention. Compared to chemical surfactants, sophorolipids are non-toxic, incredibly potent, and resistant to harsh conditions and have beneficial environmental properties (de Oliveira et al. 2015). In a shake flask culture, *Wickerhamiella domercqiae* is able to grow on rice straw holocellulose hydrolyzate and produce a significant amount of sophorolipids (53.7 g/L) (Liu et al. 2016). Using rice straw as a substrate, a heterogeneous mixture of rhamnolipids from *Achromobacter* sp. produced 31.73% of a mono-to-di-rhamnolipid ratio of 2.1:1 (Joy and Sharma 2021).

3.4.9 Biological Control Agents

Microbial fermentation techniques such as submerged and solid-state fermentation have been in demand for the development and production of novel biocontrol agents by utilizing lignocellulosic as a substrate. *B. bassiana* and *T. harzianum* were used in fixed bed reactors to study the production of fungal bioinsecticides from rice husks (Sala et al. 2020). *T. asperellum* cultured on fortified rice starch suppressed *F. oxysporum* by 66.26% (Singh et al. 2021a). Due to its high silica content, rice straw enriched with a novel source of nano-silica. The MIC (minimum inhibitory concentration) of essential clove oil against *F. oxysporum* was determined to be 125 mg/L (Singh et al. 2023). The rice plants developed resistance to the pathogen *Pyricularia oryzae*, which causes rice blast and neck rot, after treatments with the biopesticide *P. putida* H10 (Hoesain and Wahyuni 2018).

3.4.10 Substrate for Enzyme Production

To make the whole process more affordable, attention is currently being paid for developing innovative, environmentally friendly technologies to produce key enzymes for industry. The agri-residues are utilized for enzyme production as the substrate available is cost-effective and greener method (Fig. 3.5). Recently, many industries, including oleochemicals, dairy, pharmaceuticals, detergents, and

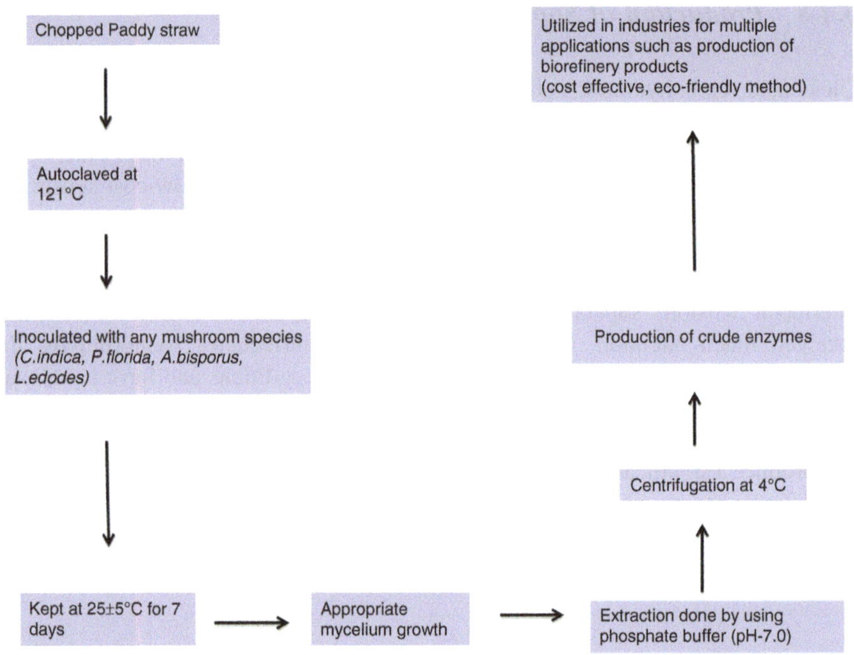

Fig. 3.5 Enzyme production using paddy straw as a substrate under solid state fermentation

agricultural products, have shown interest in enzymes as potential biocatalysts. The excessive utilization of lignocellulosic residues reduces raw material costs approximately by 60%, thus enhancing the cost-expensive process (Kaur et al. 2023). In the quest to produce cellulase from *T. reesei* and *H. insolens*, rice straw was used as the carbon source (Kogo et al. 2017). *Streptomyces psammoticus* was found to produce laccase (17.3 U/g) while using lignocellulosic biomass as a substrate (Niladevi et al. 2007). The highest levels of hydrolytic enzymes such as carboxymethylcellulose and filter paper were produced by *Phanerochaete chrysosporium* using paddy straw as the substrate (Khan et al. 2007). Using paddy straw as the substrate, *Trichoderma reesei* produced more cellulase (27.04 U/mg) during submerged fermentation (Naher et al. 2021).

3.4.11 Microbial Pigments

The detrimental effects of synthetic colors on the well-being of individuals and the ecosystem have led to the emergence of natural pigments, which have received considerable attention in research and industry. Natural colors come primarily from plants and microbes. However, due to a number of disadvantages such as seasonal volatility, year-round availability, low pigment yield/stability, and inefficient manufacturing methods, plant sources are not the first choice for pigments (Sodhi et al. 2022). Therefore, microbes are thought to be a good source of these colors. Numerous research studies have examined the development of microbial pigments from various agro-industrial substrates (Lopes and Ligabue-Braun 2021). The colored pigments produced by fungus *M. purpureus* in rice straw hydrolyzate were 8.62 U/mL and 20.91 U/mL (Liu et al. 2020). *P. sanguineus* was able to produce potent orange pigment with antibacterial properties in SSF at a reasonable cost (Meng et al. 2022). Rice straw can be utilized as a natural dye and can act as green technology. Since rice straw contains 49.3% carbon, it is black in color (Kusumastuti and Hidayati 2022). The vanillin was extracted from paddy straw with optimal concentration and yield of 0.4% and 3.96 g/g, respectively (Nurika et al. 2020).

3.4.12 Single Cell Oil (SCO)

Microorganisms produce intracellular lipid molecules known as single cell oils (microbial oils). These lipids, which resemble the lipids and fatty acids found in plants structurally, are produced in significant quantities by bacteria when nitrogen or phosphate levels are low (Ochsenreither et al. 2016). *T. fermentans* was cultured on rice straw hydrolyzate that had been detoxified and treated with sulfuric acid. It contained 40.1% lipids and had a biomass of 29.1 g/L. *Geotrichum candidum* showed highest lipid titer (5.8 g/L) and enhanced biomass concentration (15.1 g/L) in a 1:1 mixture of paddy straw hydrolyzate (Diwan and Gupta 2020). *T. mycotoxinivorans* S2 can produce 45.02% lipids by mixing glucose and xylose. The yeast was cultured with non-detoxified rice straw hydrolyzate to provide easily accessible sugars, and with conversion productivity, 5.17 g/L of lipids was produced (Sagia et al. 2020). According to Tsigie et al. (2012) *Y. lipolytica* had a maximum lipid content of 11.01 g/L in the rice bran hydrolyzate under ideal conditions.

3.5 Challenges in Value-Added Product Synthesis

Traditional practices that harm the environment and run counter to the Sustainable Development Goals include incineration, open landfill, and landfill. Modern production methods have been used effectively to produce a large amount of high-quality goods. Submerged fermentation and solid-state fermentation are two biotechnological techniques that have the potential to utilize waste and produce goods. However, before it can be successfully commercialized, a number of issues that limit the overall capacity of the method to synthesize elements and the quality of these products must be addressed.

Low product concentration, moderate complexity, and high production costs are the main challenges in submerged fermentation. In order to digest large amounts of lignocellulosic waste, solid-state fermentation, a basic waste processing approach with a straightforward process, needs to be improved in scale-up and bioreactor management (Leite et al. 2019). One of the most significant constraints that needs to be carefully considered to enable industrially feasible and ideal product development is the design of the bioreactor. Massive diffusion, heat transfer rate, and compound extraction are key requirements for an effective bioreactor (Mansour et al. 2016). Because agro-industrial waste has poor thermal conductivity and accounts for the majority of solid fermentation substrates, it can be difficult to dissipate heat, which can lead to additional cell damage (Wei et al. 2015).

A laser-like emphasis on product extraction and purification methods is required to deliver an innovative, cost-effective, and durable solution. The processing methods such as separation and purification have a major impact on the cost of essential product (Zhang et al. 2018). The sustainability of downstream processing should be improved through the accessibility of novel techniques and materials that meet end-product criteria (Pérez-Rivero et al. 2019). In this situation, the creation of continuous processing systems can lead to cost savings, optimal substrate utilization, and an improvement in product quality. Another area of concern is product quality and contamination, particularly during the recovery process.

When using lignocellulosic biomass, contamination of the end products with sugar residues, including salts and volatile compounds, is a common but significant problem that makes further processing difficult and affects product quality. Product yields are improved and further increased by using strains specifically designed to reduce the production of unwanted secondary metabolites or by optimizing media formulation. Extensive research on pathway optimization and techno-economic product properties is also required to develop a metabolic engineering-based bioprocess on an industrial scale (Fernández-Cabezón and Nikel 2020).

3.6 Conclusion and Future Perspective

Rice straw was considered as the mostly useless garbage. However, other uses for these minerals have been explored as disposal regulations have tightened. As a result, several advances have been made in agriculture, power generation, pollution management, and building materials. The diverse industrial uses for products made from rice biomass conversion and the high value-added products of these products continue to be a driving force behind the advancement of rice biomass conversion technology. Among other things, the properties of rice plants should be improved and the lignin concentration increased while reducing the silica content for thermo-chemical processing, elevated temperatures, fermentation methods used for ethanol, and the development of biorefineries. To adequately meet short-term green energy needs, scientists need to conduct more research to ensure such an abundant source of bioproducts is utilized.

References

Ahmad A, Banat F, Taher H (2020) A review on the lactic acid fermentation from low-cost renewable materials: recent developments and challenges. EnvironTechnol Innov 20:101138
Akinyemi B, Omoniyi TE, Elemile O, Arowofila O (2020) Innovative husk-crete building materials from rice chaff and modified cement mortars. Acta Technol Agric 23(2):67–72
Angoy A, Ginies C, Goupy P, Bornard I, Ginisty P, Sommier A, Valat M, Chemat F (2020) Development of a green innovative semi-industrial scale pilot combined microwave heating and centrifugal force to extract essential oils and phenolic compounds from orange peels. Innovative Food Sci Emerg Technol 61:102338
Asadi N, Alavijeh MK, Zilouei H (2017) Development of a mathematical methodology to investigate biohydrogen production from regional and national agricultural crop residues: a case study of Iran. Int J Hydrogen Energy 42(4):1989–2007
Azizah N (2019) Biotransformation of xylitol production from xylose of lignocellulose biomass using xylose reductase enzyme. J Food Life Sci 3(2):103–112
Baek SC, Kwon YJ (2007) Optimization of the pretreatment of rice straw hemicellulosic hydrolyzates for microbial production of xylitol. Biotechnol Bioprocess Eng 12:404–409
Belwal T, Chemat F, Venskutonis PR, Cravotto G, Jaiswal DK, Bhatt ID, Devkota HP, Luo Z (2020) Recent advances in scaling-up of non-conventional extraction techniques: learning from successes and failures. TrAC Trends Anal Chem 127:115895
Bevilaqua DB, Rambo MK, Rizzetti TM, Cardoso AL, Martins AF (2013) Cleaner production: levulinic acid from rice husks. J Clean Prod 47:96–101
Cañon C, Sanchez N, Cobo M (2022) Sustainable production of ethyl levulinate by levulinic acid esterification obtained from Colombian rice straw. J Clean Prod 377:134276
Celligoi MAP, Silveira VA, Hipólito A, Caretta TO, Baldo C (2020) Sophorolipids: a review on production and perspectives of application in agriculture. Spanish J Agric Res 18(3):e03R01
Chakraborty S, Uppaluri R, Das C (2020) Optimization of ultrasound-assisted extraction (UAE) process for the recovery of bioactive compounds from bitter gourd using response surface methodology (RSM). Food Bioprod Process 120:114–122
Chen C, Zhang B, Huang Q, Fu X, Liu RH (2017) Microwave-assisted extraction of polysaccharides from Moringa oleifera Lam. leaves: characterization and hypoglycemic activity. Ind Crop Prod 100:1–11
Chen C, Lu Y, Yu H, Chen Z, Tian H (2019) Influence of 4 lactic acid bacteria on the flavor profile of fermented apple juice. Food Biosci 27:30–36
Collivignarelli MC, Sorlini S, Milanese C, Illankoon WAMAN, Caccamo FM, Calatroni S (2022) Rice industry by-products as adsorbent materials for removing fluoride and arsenic from drinking water—a review. Appl Sci 12(6):3166
de Oliveira MR, Magri A, Baldo C, Camilios-Neto D, Minucelli T, Celligoi MAPC (2015) Sophorolipids A promising biosurfactant and it's applications. Int J Adv Biotechnol Res 6(2): 161–174
De Silva GS, Perera BVA (2018) Effect of waste rice husk ash (RHA) on structural, thermal and acoustic properties of fired clay bricks. J Build Eng 18:252–259
Diwan B, Gupta P (2020) A deuteromycete isolate Geotrichum candidum as oleaginous cell factory for medium-chain fatty acid-rich oils. Curr Microbiol 77:3738–3749
Durak H (2014) Bio-oil production from Glycyrrhiza glabra through supercritical fluid extraction. J Supercrit Fluids 95:373–386
Emelyanenko KA, Sanzharovsky NA, Chulkova EV, Ganne AA, Emelyanenko AM, Boinovich LB (2018) Superhydrophobic corrosion resistant coatings for copper via IR nanosecond laser processing. Mater Res Express 5(11):115001
Eş I, Khaneghah AM, Barba FJ, Saraiva JA, Sant'Ana AS, Hashemi SMB (2018) Recent advancements in lactic acid production-a review. Food Res Int 107:763–770
Fan S, Wang Y, Li Y, Wang Z, Xie Z, Tang J (2018) Removal of tetracycline from aqueous solution by biochar derived from rice straw. Environ Sci Pollut Res 25:29529–29540

Fernández-Cabezón L, Nikel PI (2020) Advanced metabolic engineering strategies for the development of sustainable microbial processes. In: Gupta VG (ed) New and future developments in microbial biotechnology and bioengineering. Elsevier, Amsterdam, pp 225–246

Freitas PA, Gonzalez-Martinez C, Chiralt A (2022) Applying ultrasound-assisted processing to obtain cellulose fibres from rice straw to be used as reinforcing agents. Innovative Food Sci Emerg Technol 76:102932

Guirimand G, Sasaki K, Inokuma K, Bamba T, Hasunuma T, Kondo A (2016) Cell surface engineering of Saccharomyces cerevisiae combined with membrane separation technology for xylitol production from rice straw hydrolysate. Appl Microbiol Biotechnol 100:3477–3487

Gunjan, Singh R, Pradhan S (2023) Valorization of byproducts produced during extraction and purification of biodiesel: a promising biofuel. In: Prasad L, Pradhan S, Naik SN (eds) Biofuel extraction techniques. Wiley, Hoboken, pp 333–366

Hoesain M, Wahyuni WS (2018) Biopesticide made from rice straw. UNEJ e-Proceeding 5:137–141

Jin X, Wang J, Dai L, Liu X, Li L, Yang Y, Cao Y, Wang W, Wu H, Guo S (2020) Flame-retardant poly (vinyl alcohol)/MXene multilayered films with outstanding electromagnetic interference shielding and thermal conductive performances. Chem Eng J 380:122475

Jittin V, Bahurudeen A, Ajinkya SD (2020) Utilisation of rice husk ash for cleaner production of different construction products. J Clean Prod 263:121578

Joy S, Sharma S (2021) Antimicrobial applications of rhamnolipid biosurfactant produced from Achromobacter sp. (PS1) isolate using lignocellulosic hydrolysate. Lett Appl Nanobiosci 11: 63–71

Kaur M, Sharma S, Sodhi HS (2022a) White rot fungus calocybe indica: incredulous factory of lignocellulolytic enzymes and their potential applicability for mushroom cultivation. Acta Aliment 51(3):424–436

Kaur S, Guleria P, Sidana A, Yadav SK (2022b) Efficient process for xylitol production from nitric acid pretreated rice straw derived pentosans by Candida tropicalis GS18. Biomass Bioenergy 166:106618

Kaur M, Gill MK, Sharma S, Kocher GS, Sodhi HS (2023) Biological pretreatment strategies for second-generation lignocellulosic biomass to enhance ethanol production. In: Srivastava N, Verma B, Mishra P (eds) Agroindustrial waste for green fuel application, Clean energy production technologies. Springer, Singapore, pp 169–203

Khan MH, Ali S, Fakhru'l-Razi A, Alam Z (2007) Use of fungi for the bioconversion of rice straw into cellulase enzyme. J Environ Sci Health B 42(4):381–386

Kim J, Han J (2021) Bio-based process for the catalytic production of ethyl levulinate from cellulose. Appl Energy 300:117430

Kiyoshi K, Furukawa M, Seyama T, Kadokura T, Nakazato A, Nakayama S (2015) Butanol production from alkali-pretreated rice straw by co-culture of clostridium thermocellum and clostridium saccharoperbutylacetonicum. Bioresour Technol 186:325–328

Kogo T, Yoshida Y, Koganei K, Matsumoto H, Watanabe T, Ogihara J, Kasumi T (2017) Production of rice straw hydrolysis enzymes by the fungi Trichoderma reesei and Humicola insolens using rice straw as a carbon source. Bioresour Technol 233:67–73

Krishania M, Kumar V, Sangwan RS (2018) Integrated approach for extraction of xylose, cellulose, lignin and silica from rice straw. Bioresour Technol Rep 1:89–93

Kusumastuti A, Hidayati MNU (2022) Utilization of ash from thermal decomposition of rice straw as a natural dye. J Adv Res Fluid Mech Therm Sci 95(1):206–212

Leite P, Silva C, Salgado JM, Belo I (2019) Simultaneous production of lignocellulolytic enzymes and extraction of antioxidant compounds by solid-state fermentation of agro-industrial wastes. Ind Crop Prod 137:315–322

Li PJ, Xia JL, Shan Y, Nie ZY (2015) Comparative study of multi-enzyme production from typical agro-industrial residues and ultrasound-assisted extraction of crude enzyme in fermentation with Aspergillus japonicus PJ01. Bioprocess Biosyst Eng 38:2013–2022

Li S, Zhou Y, Luo Z, Cui Y, Xu Y, Lin L, Zhao M, Guo Y, Pang Z (2018) Dual function of ammonium acetate in acetone-butanol-ethanol fermentation by clostridium acetobutylicum. Bioresour Technol 267:319–325

Liu XG, Ma XJ, Yao RS, Pan CY, He HB (2016) Sophorolipids production from rice straw via SO 3 micro-thermal explosion by Wickerhamiella domercqiae var. sophorolipid CGMCC 1576. AMB Express 6:1–11

Liu J, Luo Y, Guo T, Tang C, Chai X, Zhao W, Bai J, Lin Q (2020) Cost-effective pigment production by Monascus purpureus using rice straw hydrolysate as substrate in submerged fermentation. J Biosci Bioeng 129(2):229–236

Lizcano SC, Dávila JA, Hernández V (2019) Fruit agroindustrial wastes for preparing beverages for medicinal purposes by supercritical fluid extraction technology: Andes Berry (Rubus glaucus benth) case. In: Grumezescu AM, Holban AM (eds) Production and management of beverages. Woodhead Publishing, Cambridge, pp 151–177

Lopes FC, Ligabue-Braun R (2021) Agro-industrial residues: eco-friendly and inexpensive substrates for microbial pigments production. Front Sustain Food Syst 5:589414

Ma H, Liu WW, Chen X, Wu YJ, Yu ZL (2009) Enhanced enzymatic saccharification of rice straw by microwave pretreatment. Bioresour Technol 100(3):1279–1284

Mansour AA, Arnaud T, Lu-Chau TA, Fdz-Polanco M, Moreira MT, Rivero JAC (2016) Review of solid state fermentation for lignocellulolytic enzyme production: challenges for environmental applications. Rev Environ Sci Biotechnol 15:31–46

Meng D, Shao X, Luo SP, Tian QP, Liao XR (2022) Pigment production by a newly isolated strain Pycnoporus sanguineus SYBC-L7 in solid-state fermentation. Front Microbiol 13:1015913

Moirangthem K, Ramakrishna P, Amer MH, Tucker GA (2021) Bioactivity and anthocyanin content of microwave-assisted subcritical water extracts of Manipur black rice (Chakhao) bran and straw. Future Foods 3:100030

Moonmoon M, Uddin MN, Ahmed S, Shelly NJ, Khan MA (2010) Cultivation of different strains of king oyster mushroom (Pleurotus eryngii) on saw dust and rice straw in Bangladesh. Saudi J Biol Sci 17(4):341–345

Moreira RC, de Melo RPF, Martínez J, Marostica Junior MR, Pastore GM, Zorn H, Bicas JL (2023) Supercritical CO_2 as a valuable tool for aroma technology. J Agric Food Chem 71(24): 9201–9212

Naher L, Fatin SN, Sheikh MAH, Azeez LA, Siddiquee S, Zain NM, Karim SMR (2021) Cellulase enzyme production from filamentous fungi Trichoderma reesei and Aspergillus awamori in submerged fermentation with rice straw. J Fungi 7(10):868

Nandal P, Sharma S, Arora A (2020) Bioprospecting non-conventional yeasts for ethanol production from rice straw hydrolysate and their inhibitor tolerance. Renew Energy 147:1694–1703

Ng HS, Kee PE, Yim HS, Chen PT, Wei YH, Lan JC (2020) Recent advances on the sustainable approaches for conversion and reutilization of food wastes to valuable bioproducts. Bioresour Technol 302:122889

Niladevi KN, Sukumaran RK, Prema P (2007) Utilization of rice straw for laccase production by streptomyces psammoticus in solid-state fermentation. J Ind Microbiol Biotechnol 34(10): 665–674

Nurika I, Suhartini S, Azizah N, Barker GC (2020) Extraction of vanillin following bioconversion of rice straw and its optimization by response surface methodology. Molecules 25(24):6031

Ochsenreither K, Glück C, Stressler T, Fischer L, Syldatk C (2016) Production strategies and applications of microbial single cell oils. Front Microbiol 7:1539

Ortiz FG, de Santa-Ana P (2017) Techno-economic assessment of an energy self-sufficient process to produce biodiesel under supercritical conditions. J Supercrit Fluids 128:349–358

Pandey A, Kumar B (2020) Investigation on the effects of acidic environment and accelerated carbonation on concrete admixed with rice straw ash and microsilica. J Build Eng 29:101125

Patil PD, Yadav GD (2018) Application of microwave assisted three phase partitioning method for purification of laccase from Trametes hirsuta. Process Biochem 65:220–227

Pérez-Rivero C, López-Gómez JP, Roy I (2019) A sustainable approach for the downstream processing of bacterial polyhydroxyalkanoates: state-of-the-art and latest developments. Biochem Eng J 150:107283

Pinotti LM, Paulino LB, Agnezi JC, dos Santos PA, da Silva HN, Zavarise JP, Salom GSB, Tardioli PW (2020) Evaluation of different fungi and bacteria strains for production of cellulases by submerged fermentation using sugarcane bagasse as carbon source: effect of substrate concentration and cultivation temperature. Afr J Biotechnol 19(9):625–635

Qi B, Yao R (2007) L-lactic acid production from lactobacillus casei by solid state fermentation using rice straw. Bioresources 2(3):419–429

Rathour RK, Devi M, Dahiya P, Sharma N, Kaushik N, Kumari D, Kumar P, Baadhe RR, Walia A, Bhatt AK, Bhatia RK (2023) Recent trends, opportunities and challenges in sustainable management of rice straw waste biomass for green biorefinery. Energies 16(3):1429

Ring LC, Yenn TW, Nazri NIB, Nee TW, Zahan KA, Ibrahim D, Rashid S (2018) Chemical optimization of red pigment, Monascorubin production in Penicillium minioluteum ED24 using solid-state fermentation. Arab J Sci Eng 43:3485–3491

Roberto IC, Mussatto SI, Rodrigues RC (2003) Dilute-acid hydrolysis for optimization of xylose recovery from rice straw in a semi-pilot reactor. Ind Crop Prod 17(3):171–176

Robles-Jimarez HR, Sanjuan-Navarro L, Jornet-Martínez N, Primaz CT, Teruel-Juanes R, Molins-Legua C, Ribes-Greus A, Campíns-Falcó P (2022) New silica based adsorbent material from rice straw and its in-flow application to nitrate reduction in waters: process sustainability and scale-up possibilities. Sci Total Environ 805:150317

Sagia S, Sharma A, Singh S, Chaturvedi S, Nain PKS, Nain L (2020) Single cell oil production by a novel yeast Trichosporon mycotoxinivorans for complete and ecofriendly valorization of paddy straw. Electron J Biotechnol 44:60–68

Sala A, Artola A, Sánchez A, Barrena R (2020) Rice husk as a source for fungal biopesticide production by solid-state fermentation using B. Bassiana and T. Harzianum. Bioresour Technol 296:122322

Salimi M, Safari F, Tavasoli A, Shakeri A (2016) Hydrothermal gasification of different agricultural wastes in supercritical water media for hydrogen production: a comparative study. Int J Ind Chem 7:277–285

Sangon S, Hunt AJ, Ngernyen Y, Youngme S, Supanchaiyamat N (2021) Rice straw-derived highly mesoporous carbon-zinc oxide nanocomposites as high performance photocatalytic adsorbents for toxic dyes. J Cleaner Prod 318:128583

Satpute SK, Płaza GA, Banpurkar AG (2017) Biosurfactants' production from renewable natural resources: example of innovativeand smart technology in circular bioeconomy. Manage Syst Prod Eng 25(1):46–54

Sik B, Hanczné EL, Kapcsándi V, Ajtony Z (2020) Conventional and nonconventional extraction techniques for optimal extraction processes of rosmarinic acid from six Lamiaceae plants as determined by HPLC-DAD measurement. J Pharm Biomed Anal 184:113173

Singh R, Patel M (2022) Effective utilization of rice straw in value-added by-products: a systematic review of state of art and future perspectives. Biomass Bioenergy 159:106411

Singh R, Tiwari S, Srivastava M, Shukla A (2014) Microwave assisted alkali pretreatment of rice straw for enhancing enzymatic digestibility. J Energy 2014:1–7

Singh B, Thakur V, Bhatia G, Verma D, Singh K (2016) Eco-friendly and cost-effective use of rice straw in the form of fixed bed column to remove water pollutants. J Bioremed Biodegr 7:374

Singh G, Tiwari A, Gupta A, Kumar A, Hariprasad P, Sharma S (2021a) Bioformulation development via valorizing silica-rich spent mushroom substrate with Trichoderma asperellum for plant nutrient and disease management. J Environ Manag 297:113278

Singh S, Kaur D, Yadav SK, Krishania M (2021b) Process scale-up of an efficient acid-catalyzed steam pretreatment of rice straw for xylitol production by C. Tropicalis MTCC 6192. Bioresour Technol 320:124422

Singh G, Arora H, Hariprasad P, Sharma S (2023) Development of clove oil based nanoencapsulated biopesticide employing mesoporous nanosilica synthesized from paddy straw via bioinspired sol-gel route. Environ Res 220:115208

Sodhi AS, Sharma N, Bhatia S, Verma A, Soni S, Batra N (2022) Insights on sustainable approaches for production and applications of value added products. Chemosphere 286:131623

Soorbaghi FP, Isanejad M, Salatin S, Ghorbani M, Jafari S, Derakhshankhah H (2019) Bioaerogels: synthesis approaches, cellular uptake, and the biomedical applications. Biomed Pharmacother 111:964–975

Soquetta MB, Terra LDM, Bastos CP (2018) Green technologies for the extraction of bioactive compounds in fruits and vegetables. CyTA-J Food 16(1):400–412

Tsigie YA, Wang CY, Kasim NS, Diem QD, Huynh LH, Ho QP, Truong CT, Ju YH (2012) Oil production from Yarrowia lipolytica Po1g using rice bran hydrolysate. J Biomed Biotechnol 2012:378384

Ukarde TM, Pawar HS (2022) PolyE-IL, an efficient and recyclable bronsted acid catalyst for conversion of rice straw into levulinic and other organic acids. Energy Fuel 36(3):1592–1603

Veena SS, Pandey M (2011) Paddy straw as a substrate for the cultivation of Lingzhi or Reishi medicinal mushroom, Ganoderma lucidum (W. Curt.: Fr.) P. Karst. in India. Int J Med Mushrooms 13(4):397–400

Vu D, Thi N, Tran H, Bui ND, Vu C, Viet NH (2017) Lignin and cellulose extraction from Vietnam's rice straw using ultrasound-assisted alkaline treatment method. Int J Polym Sci 2017: 1063695

Wei K, Lv C, Chen M, Zhou X, Dai Z, Shen D (2015) Development and performance evaluation of a new thermal insulation material from rice straw using high frequency hot-pressing. Energ Buildings 87:116–122

Wu Y, Fan Y, Zhang M, Ming Z, Yang S, Arkin A, Fang P (2016) Functionalized agricultural biomass as a low-cost adsorbent: utilization of rice straw incorporated with amine groups for the adsorption of Cr (VI) and Ni (II) from single and binary systems. Biochem Eng J 105:27–35

Yaaini N, Amin NAS, Asmadi M (2012) Optimization of levulinic acid from lignocellulosic biomass using a new hybrid catalyst. Bioresour Technol 116:58–65

Yang CY, Fang TJ (2014) Combination of ultrasonic irradiation with ionic liquid pretreatment for enzymatic hydrolysis of rice straw. Bioresour Technol 164:198–202

Yang T, Wang J, Li B, Kai X, Xing W, Li R (2018) Behaviors of rice straw two-step liquefaction with sub/supercritical ethanol in carbon dioxide atmosphere. Bioresour Technol 258:287–294

Yao R, Qi B, Deng S, Liu N, Peng S, Cui Q (2007) Use of surfactants in enzymatic hydrolysis of rice straw and lactic acid production from rice straw by simultaneous saccharification and fermentation. Bioresources 2(3):389–398

Zaman M, Hamid B (2022) 7 biosurfactants production and applications in food. In: Sayyed RZ, El-Enshasy H (eds) Microbial surfactants, Applications in food and agriculture, vol 2. CRC Press, Boca Raton, pp 133–149

Zhang QW, Lin LG, Ye WC (2018) Techniques for extraction and isolation of natural products: a comprehensive review. Chin Med 13:20

Zhu S, Huang W, Huang W, Wang K, Chen Q, Wu Y (2015) Pretreatment of rice straw for ethanol production by a two-step process using dilute sulfuric acid and sulfomethylation reagent. Appl Energy 154:190–196

Zikriyani H, Saskiawan I, Mangunwardoyo W (2018) Utilization of agricultural waste for cultivation of paddy straw mushrooms (Volvariella volvacea (bull.) singer 1951). Intl J Agric Technol 14(5):805–814

Chapter 4
Valorization of Paddy Straw Waste for Sustainable Development of Biofuels

Manmeet Kaur, Ravneet Kaur, Shivani Sharma, and G. S. Kocher

Abstract Growing rice, an important global staple, generates a substantial amount of straw-based biomass waste. Rice straw deserves special attention as a biomass substrate for the development and production of next-generation biofuels due to its wide availability and solid lignocellulose composition. Lignocellulosic biomass, which is environmentally friendly, inexpensive, and renewable, can be converted into biofuels. Agricultural waste can be processed into biofuel to improve waste disposal, stop environmental deterioration, and ensure reliability in energy. The methods of using rice straw as a lignocellulosic substrate for the generation and production of biofuels such as bioethanol, biohydrogen, biobutanol, biogas, and biomethanol form the basis for the content of this chapter. In addition, the importance of pretreatment techniques to break down the complex lignocellulosic biomass and the implementation of fermentation to convert reducing sugars into biofuel was highlighted.

Keywords Paddy straw · Lignocellulosic biomass · Bioethanol · Biobutanol · Biohydrogen · Biogas and biomethanol · Sustainable environment

4.1 Introduction

Numerous challenges, including rising fuel prices, increased fuel consumption, worrying global warming, and the depletion of renewable sources, have motivated the search for viable alternatives to energy production. However, energy consumption will increase maximally by 1.3% in the coming next years. Therefore, all countries must change their energy policies without delay to ensure global energy security. Several methods for converting biomass into biofuels have recently been developed, taking into account their transport, storage, and energy densities (Vamvuka 2011). The biomass content and optimal reaction parameters such as

M. Kaur (✉) · R. Kaur · S. Sharma · G. S. Kocher
Department of Microbiology, Punjab Agricultural University, Ludhiana, Punjab, India
e-mail: manmeet1-mb@pau.edu

N. Srivastava et al. (eds.), *Paddy Straw Waste for Biorefinery Applications*, Clean Energy Production Technologies, https://doi.org/10.1007/978-981-99-8224-0_4

feed sizes, temperature conditions, reaction speed, and heating rate have an impact on the biofuel yield. Biofuels, which are now an intriguing option to reduce greenhouse gas emissions and combat biodiversity loss, support the idea of sustainable development, which includes improved food production from biomass and effective use of water and many other resources (Sindhu et al. 2019). Agricultural residues can be fermented to produce numerous biofuels such as biogas, biomethanol, biohydrogen, biodiesel, and bioethanol (Singh and Chaudhary 2017). Due to its accessibility and widespread use, rice straw is a promising substrate for bioenergy production. It is one of the main crops that produce enormous amounts of silica-enriched biomass (Satlewal et al. 2017). The hazardous gases which are released during combustion increase pollution, public health problems, and greenhouse gas emissions (Wang et al. 2016). The durability of rice straw as a fuel source that enables conscientious bioenergy production is one of the main obstacles. Rice straw requires a suitable pretreatment method due to its chemical structure consisting of cellulose, lignin, and hemicellulose (Agrawal et al. 2017). Silica which is abundant in rice and is necessary for the mechanical strength of cell walls also represents its significant importance. However, it is not known how this affects the fermentation and saccharification processes. Consequently, rice straw consumption poses a significant challenge, and it is crucial to improve the financial viability of such conversion technologies (Huang and Lo 2019).

4.2 Paddy Straw-Lignocellulosic Biomass

The most important cereal crop, rice straw (*Oryza sativa*), produces a lot of agricultural waste in the form of excess straw. Although this residual biomass is often used as a raw material by numerous industries, significant parts of the residues are incinerated or landfilled (Sharma et al. 2020). When the straw is ruthlessly burned, numerous pollutants such as polycyclic aromatic hydrocarbons (n-alkanes), carbon-dioxide, traces of sulfur dioxide, and carbon monoxide are released into the environment. These pollutants are linked to serious human diseases such as asthma and cancer. Carbon dioxide emissions from open-air combustion also pose a major threat to the ecosystem and also had a major impact on global warming. Meanwhile, the widespread practice of discarding rice straw in the field has resulted in significant environmental damage and financial loss (Pham et al. 2019).

4.3 Structure of Agro-industrial Residue

It is important to understand the structure of lignin, especially its chemical composition, as well as the biosynthesis process in order to develop a pretreatment method that can soften straw and add hydrolytic enzymes such as cellulase for cellulose depolymerization. Figure 4.1 shows a typical conceptual framework for agro-

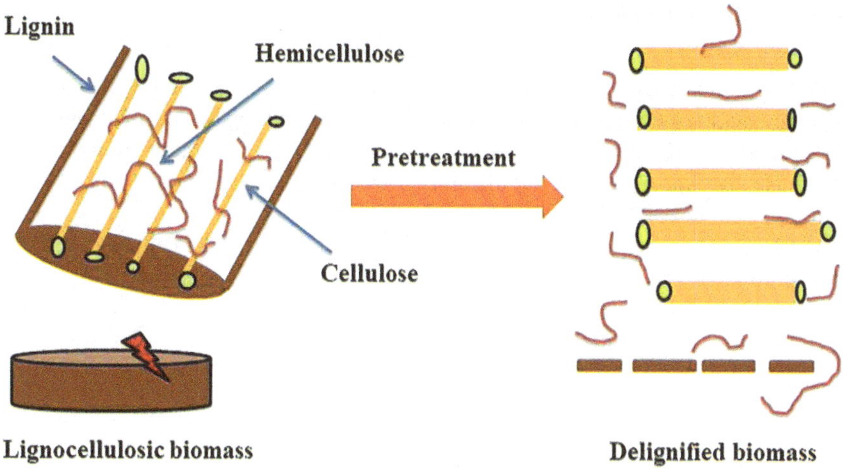

Fig. 4.1 Schematic representation of pretreatment impact on lignocellulosic biomass (Adapted from Nair and Sivakumar 2020)

industrial biomass. The cellulose microfibrils are tightly attached to the hemicellulose matrix during cell formation, as they first divide to develop the basic cell wall. The formation of connective tissue in the body occurs through the middle lamella, a pectin-based layer, which fuses with the main wall of the cell to form the conductive tissue system. Meanwhile, cell development concludes that the primary wall, secondary wall, and plasma membrane form the two upper layers of the cell. By concentrating lignin in xylem fibers, the secondary wall often improves structure by increasing mechanical strength. The majority of the biomass or xylem fibers can be converted into bioethanol. To increase the stability of lignocellulose and prevent its degradation, as well as to promote nutrition and water transport, the conductive tissue system cross-links cellulose, hemicelluloses, and lignin (Hans et al. 2019).

4.3.1 Cellulose

The most common lignocellulosic biomass polymer, cellulose, is made up of cellobiose as the main repeat unit which are connected by glycosidic linkages. The microfibrilized cellulose chains consisting of glucose units are arranged into cellulose fibrils by being bundled together (Robak and Balcerek 2018). Due to the incorporation of cellulose fibrils into a lignocellulosic matrix, it is particularly resistant to enzymatic hydrolysis. These hydrogen bonds align the chains parallel to each other, forming a crystalline structure. As a result, cellulose microfibrils have sections that are both highly crystalline and less ordered amorphous (Yoo et al. 2017).

4.3.2 Hemicellulose

20–35% of the weight of biomass consists of hemicelluloses, heterogeneous groupings of biopolymers (Chandel et al. 2018). Xylans, xyloglucans, mannans, and glucomannans are formed from a variety of monosaccharide subunits found within them. Hemicelluloses have a significantly lower degree of polymerization as compared to cellulose. The cellulose, lignin, and hemicellulose network as a whole is more rigid due to the presence of hemicellulose that connects the cellulose fibers to the lignin (Tian et al. 2018).

4.3.3 Lignin

After cellulose, lignin is the second-most prevalent polymer; it makes up roughly 12–38% of the dry weight of lignocellulosic biomass (Ragauskas et al. 2014). The phenylpropanoid building blocks make up this extremely complex amorphous heteropolymer. Properties of lignin include structural rigidity and hydrophobicity. Due to the role of lignin as a barrier that prevents enzymes from penetrating cellulose, the physical availability of macromolecules may be limited (Zoghlami and Paes 2019). Furthermore, due to its hydrophobic structural properties, it is capable of adsorbing cellulases and other cellulolytic enzymes during enzymatic hydrolysis (Zeng et al. 2014). Lignin elimination often disrupts the lignin-carbohydrate matrix and increases permeability and removal of enzyme adsorption sites (Kruyeniski et al. 2019).

4.4 Pretreatment Methods of Rice Straw

Paddy straw can be used on various platforms to produce biofuels and bioenergy. Before rice straw can be used on a biological platform, it must undergo a pretreatment phase that alters the relationship between major components (cellulose, hemicellulose, and lignin). This alteration enhanced cellulose accessibility, removes carbohydrate-lignin complex, and decreased the cellulose crystallinity. The multiple pretreatment methods increase the biofuel production via enzymatic hydrolysis of the lignocellulosic biomass (Kaur et al. 2023). Table 4.1 lists the pretreatment techniques used to convert lignocellulosic biomass into contributory biofuels (Kumar and Sharma 2017).

Table 4.1 Pretreatment methods of lignocellulosic biomass (Kumar and Sharma 2017)

Pretreatment method	Type of pretreatment technologies	Advantages	Disadvantages
Physical pretreatment	Mechanical treatment	Reduction in cellulose crystallinity and often leads to minimum degree of polymerization	Maximum energy output
	Steam pretreatment	Hemicellulose breakdown resulted in degradation	Maximum temperature and pressure are mandatory
Chemical pretreatment	Acid pretreatment	Hydrolysis of hemicellulose; enzyme digestibility occurred through cellulose	Corroded substances
	Alkali pretreatment	Eliminates lignin	Minimum effective in lignin related biomass
Physicochemical pretreatment	Ozone pretreatment	Lignin content decreased to some extent	Highly priced
	Organosolv process	Breaks down the complex bond between hemicellulose and lignin	Elimination of solvents
	Steam explosion	Alteration in the complex structure leads to the degradation of carbon bond between hemicellulose and lignin	Difficult to utilize in softwoods as low amount of acetyl molecule is found in the hemicellulose component
	Ammonia free expansion	Enzymatic hydrolysis generates destruction of cell wall	The biomass with high lignin content is not much used in this process
Biological pretreatment	Biological pretreatment (white rot fungi and lignolytic microorganisms)	Delignification	Prolonged incubation time
	Ligninolytic enzymes (in vitro degradation)	Alteration of lignin molecule often leads to degradation	Hydrolysis rate is quite low, and secondly it is the most expensive
	Consolidated bioprocessing	Cost-effective	Urgency of both cellulolytic and alcoholic microorganisms

4.5 Biofuels from Paddy Straw

4.5.1 Bioethanol

Bioethanol is produced by fermenting natural resources, typically plant derivatives or agricultural waste, to yield a liquid, sustainable biofuel. The primary driver of increased output was the need for bioethanol. Figure 4.2 shows the representation of

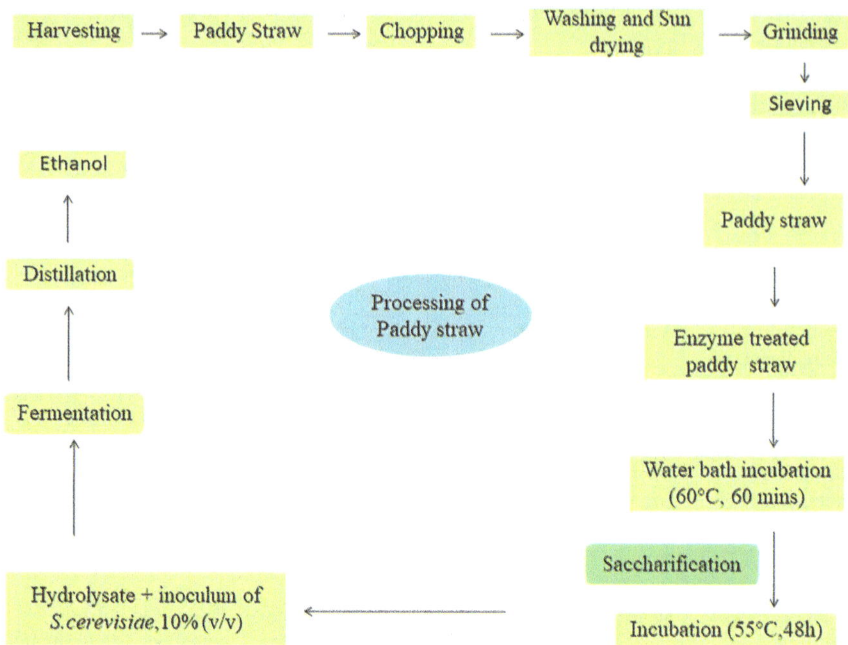

Fig. 4.2 Schematic representation of enzymatic hydrolysis of paddy straw for bioethanol production

the enzymatic hydrolysis of paddy straw to produce bioethanol. In 2016, the world produced 100.2 L of bioethanol. By 2024, there will be 144.5 billion liters of bioethanol in use (Bušić et al. 2018). It is primarily used to fuel internal combustion engines. Bioethanol production from rice husks and rice straw resulted in maximum yield of 44.02% and 18.05 g/L, respectively (Srivastava et al. 2014). Rice straw was saccharified with a sodium hydroxide (1%) yielding 22.15 g/L of reducing sugars. *S. tannophilus* was inoculated in the substrate which leads to the ethanol production with a yield and concentration of 9.53 g/L and 84.15%, respectively (Jin et al. 2020). The fermentable sugar content of rice straw after pretreatment with alkali increased from 56.3% to 80.0%. The optimized cocktail hydrolyzed alkali-treated paddy straw with an efficiency of approximately 94.3% and produced a total of 75.3 g/L fermentable sugars. After exposure to rice straw treated with 100 g/L SSF and the ideal cocktail, *M. circinelloides* produced 31.03 g/L ethanol under aerobic conditions (Takano and Hoshino 2018). *S. cerevisiae* MTCC-173 produced ethanol and sugar yield (g/l) of 6.71 and 27.42 while simultaneously delignifying, saccharifying, and fermenting rice straw in a single step (Bhardwaj et al. 2020). In cellulase hydrolysis of rice straw, bacterial laccase produces glucose (16.1 mg/mL), reducing sugars (21.82 mg/mL), and bioethanol (22.11 mg/mL) (Nazar et al. 2022).

4.5.2 Biobutanol

Biobutanol is believed to have better fuel properties due to its greater energy content and improved air-fuel ratio, which make it a more efficient fuel. Due to its higher ignition point and lower vapor pressure than bioethanol, handling biobutanol is less dangerous. Instead of 10% ethanol, fuel can contain up to 15% isobutanol, a branched isomer of butanol. However, butanol is comparable to gasoline, although it has a lower octane number than ethanol. So butanol cannot be used to increase the octane number. The toxicity of butanol appears to be its main disadvantage (Liu et al. 2022). When microorganisms ferment renewable raw materials such as sugar, starch, or cellulose, butanol with four carbon atoms is produced. It is an advanced biofuel that vaporizes at higher temperature. Due to low fermentation yield, biofouling, and scaling issues, the synthesis of biobutanol is not commercially viable (Karthick and Nanthagopal 2021). Biobutanol yields of rice straw and sugarcane baggase of 2.90 and 1.91 g/L were achieved (Cheng et al. 2012). Researchers investigated that the anaerobic fermentation of rice straw with *C. acetobutylicum* resulted in a butanol yield of 14.12 g/L. Paddy straw was prepared by acidic enzymatic hydrolysis with butanol production (3.43 g/L) (Gottumukkala et al. 2013). *Pichia* sp. and *S. cerevisiae* produced 4.32 g/L of butanol (Mohapatra et al. 2020). Fermentation of rice straw hydrolyzate by *C. acetobutylicum* resulted in production of biobutanol as 0.27 g/g glucose (Rahnama et al. 2014). Delignified paddy straw was inoculated with *C. thermocellum* and *C. saccharoperbutylacetonicum* hence resulting in biobutanol production of approximately around 5.5 g/L (Kiyoshi et al. 2015). Rice straw fermented with *C. beijerinckii* produced more biobutanol (8%) under solid-state fermentation as compared to separate hydrolysis and fermentation process (Valles et al. 2020).

4.5.3 Biohydrogen

Biohydrogen is a cutting-edge biofuel produced from biomass utilizing thermochemical and biological methods. It is secure, non-toxic, and devoid of carbon. It is a highly flammable, flavorless, odorless, and environmentally friendly renewable fuel. Due to its superior energy content and calorific value, it is the most beneficial and best alternative among all biofuels. The use of biohydrogen, which has multiple uses in industry, power generation, the manufacture of food and beverages, transportation, and healthcare, substantially improves the capacitance (Kumari and Singh 2018). Due to the use of accessible, sustainable and highly renewable lignocellulosic biomass, biohydrogen has an advantage over other renewable fuels (Awogbemi and Von Kallon 2022). The two most commonly used methods to produce biohydrogen are fermentation and biophotolysis. Biohydrogen production can also be accomplished by pyrolysis, solar gasification, thermochemical gasification, and supercritical conversion. These manufacturing approaches are more environmentally friendly, economically advantageous, and sustainable compared to traditional methods (Singh et al. 2015).

Hydrogen can be produced using numerous advanced biohydrogen technologies, including direct and indirect biophotolysis and photo-dark fermentations. Biochemical and electrochemical methods of hydrogen synthesis are superior from an ecological and energetic perspective. Cyanobacteria convert water into hydrogen and oxygen via photosynthesis. Anaerobic bacteria can only obtain their electrons and energy from organic molecules from which they produce hydrogen. Bacteria such as *Clostridia* can produce biohydrogen by changing the temperature, pH, hydraulic residence time (HRT) of the reactor, and other elements of the treatment system. By using organisms such as microalgae, the fermentation process, and photodegradation of organic materials, biohydrogen can be developed from agricultural crops and photosynthetic bacteria (Hussy et al. 2005). To produce hydrogen gas, microorganisms such as *Aerobacter*, *Escherichia coli*, and *Clostridium* spp. ferment butyric acid, butylene glycol, and mixed acids. The dilution and metabolites produced during fermentation were studied using this method. Anaerobic sludge also utilizes heat shock to increase the hydrogen fermentation productivity of the substrate (Keskin and Emiroglu 2010). *Thermoanaerobacterium thermosaccharolyticum* M18 was added to the substrate after being treated with urea and maximum biohydrogen production resulted to be 22.07 mmol/L (Dong et al. 2018). Asadi and Zilouei (2017) reported that organosolv treatment of rice straw resulted in 19.73 mL/g hydrogen. Rice straw hydrolyzate produced 5.52 L/L/day of biohydrogen in a bioreactor with continuous external circulation (Liu et al. 2014). After exposure to 1.0% sulfuric acid, rice straw produced 55.6 mL of hydrogen (Kim et al. 2022). During the treatment of rice straw with sulfuric acid, hydrogen production rate enhanced to 250 mL/L/h (Safari et al. 2022). After pretreatment of rice straw with *Gymnopus contrarius* J2, the hydrogen produced increased to 5.71 mmol/g (Sheng et al. 2018). During *Proteus mirabilis* inoculation, the rice straw hydrolysates treated with acid and alkali can produce hydrogen. However, the alkaline hydrolyzate produced less hydrogen than the acidic hydrolyzate (Mechery et al. 2021). According to Sen et al. (2016), the hydrolyzate of already treated rice straw produced a lot of hydrogen at a rate of 771 mL/L. Researchers found that cellulose waste contains about 3–4.5 moles of the H_2/glucose molecule (Phillips 2021).

4.5.4 Biogas

Biogas, a valuable fuel, is produced as a raw product in fermenters from wastewater or compost waste. Organic products are difficult to break down by anaerobic bacteria. Organic chemicals accumulate when there is a lack of oxygen and release methane. Anaerobic bacteria undergo a series of metabolic reactions to break down organic molecules into simple compounds (Fig. 4.3). Anaerobic bacteria that produce methane require organic acids to complete the degradation process. Acidogens

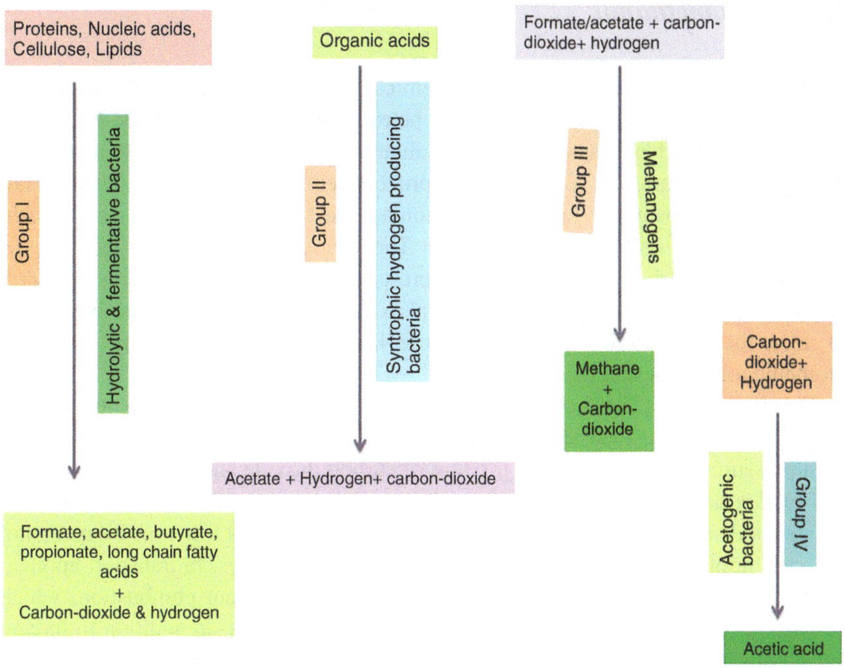

Fig. 4.3 Microorganisms and chemical reactions which are involved during anaerobic digestion

convert the unprocessed waste into less complicated fatty acids in the very first stage. Methanogens assimilate the organic acids produced in the second stage by acidogens, resulting in the synthesis of biogas. In the final step, the acids of the organisms are completely converted into biogas (Kougias and Angelidaki 2018). Simultaneous digestion of microalgae biomass (30%) and rice straw (70%) accelerated the biogas production (Dar et al. 2019). Researchers studied that rice straw treated with *C. versicolor* improved the amount of biogas rate by 26.01% (Phutela et al. 2011). Rice straw processed with sodium carbonate in the microwave resulted in a 54.4% increase in biogas production (Kaur and Phutela 2016). The highest biogas value was determined with a methane content of 58.3% after 35 days (Pal et al. 2022). Wati and Narula (2017) found that adding rice straw to biogas plants increased gas production from 2.76% to 16.32%. When rice straw was treated with sodium hydroxide, a biogas rate of 49.7% was produced (Garg et al. 2017). Rice straw treated with *Paecilomyces*, *Aspergillus*, and *Penicillium* increased biogas production by 41.5%, 37.01%, and 25.02%, respectively (Sahni and Phutella 2017). The largest biogas yield (360.8 mL/g) and the maximum biogas production rate (8.17 mL/g/day) were obtained when the biomass and rice straw were simultaneously digested in a 1:1 ratio (Dar et al. 2021).

4.5.5 Biomethanol

Methanol, one of the most significant chemicals in industry, can be used as a clean fuel or added to petrol. For conversion to petrol, a shape-selective (ZSM-5) catalyst can be utilized. There hasn't been a sustainable, commercially feasible method of producing methanol yet. The chemical process required to create methanol from biomass is too costly. So far, the only biomass used to produce methanol is trash, such as waste wood or organic waste. The biomass is converted into syngas (hydrogen and carbon monoxide) in the presence of oxygen. Methanol is currently produced using natural gas; however partial oxidation techniques can also be utilized to create the chemical from biomass. Figure 4.4 displays the primary methanol and biomethanol production facilities.

Demirbas et al. (2011) claim that in order to create biomethanol, biomass must first be routinely gasified at high temperature, followed by a catalytic synthesis at high pressures from the resultant mixture of carbon dioxide and hydrogen. Biomass and biodegradable waste with sugar and starch are converted into biomethanol, an oxygenated sustainable fuel. Due to its high octane rating and low cost, biomethanol is a commonly available fuel for internal combustion engines. In transport applications, the use of biomethanol ensures complete and efficient combustion, which ultimately results in a decrease in greenhouse gas emissions. In addition to having excellent performance, biomethanol required minimum temperature, medium flammability, and nitric oxide emissions. Furthermore, it also has a low energy content. Additionally, when biomethanol is burned inefficiently, formaldehyde, formic acid, and other pollutants are emitted (Gautam et al. 2020). Methanol is the alcohol of choice for the trans-esterification process in addition to being used as a transportation fuel because of its affordability, simplicity of recovery, superior reactivity, and propensity to create azeotropes. Methanol is a key ingredient in the synthesis of acetic acid, formaldehyde, chloromethane, and other chemical molecules.

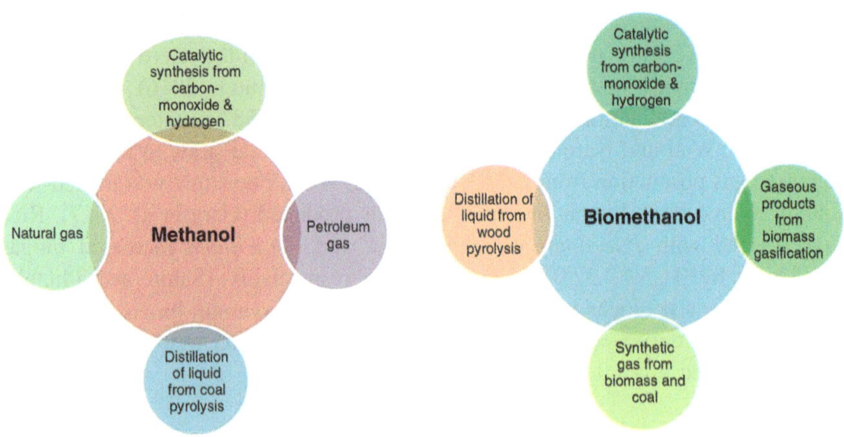

Fig. 4.4 Major production facilities of methanol and biomethanol

Pharmaceuticals, inks, synthetic resins, adhesives, coatings, and automobile anti-freeze are all made using methanol (Sarangi et al. 2020). Researchers found that rice straw can be turned into biomethanol with a lower yield and an energy efficiency (40.5%). Because of its low cost, ease of recovery, superior reactivity, and propensity to produce azeotropes, methanol is the alcohol of choice for the trans-esterification process in addition to being utilized as a transportation fuel (Xiao et al. 2009).

4.6 Merits and Demerits of Producing Biofuels from Agro-industrial Residue

There is a discussion about fossil fuels that focuses on the pros and cons of biofuels produced from agricultural byproducts, including reductions in greenhouse gas emissions, greater efficiency, and greater biodiversity loss. According to Barisano et al. (2020), biofuels have the following advantages and disadvantages (Fig. 4.5).

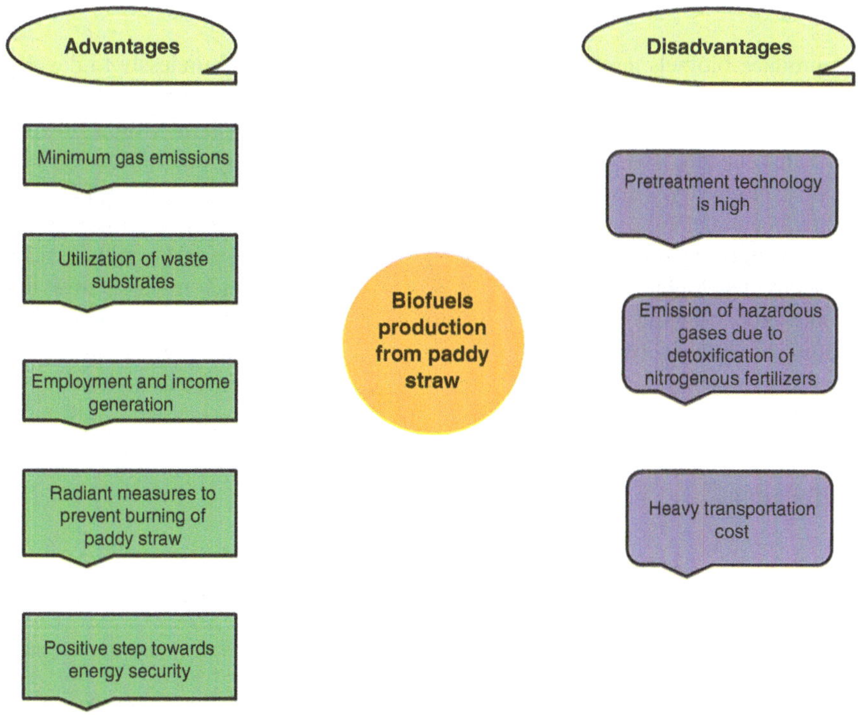

Fig. 4.5 Positive and negative impact of biofuel production from paddy straw

4.6.1 *Merits*

- Reducing greenhouse gas emissions
- Production of renewable energy
- Reducing energy imports to energy-scarce countries
- Improvement of the economic structure in rural regions
- Replacing petrochemical companies with those based on renewable resources

4.6.2 *Demerits*

- Cellulosic biofuels eliminate crop residues which are critical to maintaining soil health.
- Loss of biodiversity.
- Technological advances to take advantage of soil.
- The need for financial incentives for economic viability.

4.7 Conclusion

To produce biofuels and bioenergy, researchers have worked tirelessly to develop innovative bioconversion technologies, taking into account the harmful effects that the widespread availability of large quantities of rice straw will have on nature and individual health. The composition of paddy straw makes it difficult to use it to produce bioenergy. This resilient composition allows biorefineries to use appropriate pretreatment methods, increasing costs and complicating economic analysis. Several research and development initiatives are underway around the world to address these issues by developing novel useful technologies. The three main research areas that should be considered by all research sectors are (a) reduced economic impact on biofuel production, (b) increased efficiency, and (c) reduced or no formation of dangerous gases or toxins and consequently no harm to the entire ecosystem.

References

Agrawal R, Satlewal A, Kapoor M, Mondal S, Basu B (2017) Investigating the enzyme-lignin binding with surfactants for improved saccharification of pilot scale pretreated wheat straw. Bioresour Technol 224:411–418

Asadi N, Zilouei H (2017) Optimization of organosolv pretreatment of rice straw for enhanced biohydrogen production using Enterobacter aerogenes. Bioresour Technol 227:335–344

Awogbemi O, Von Kallon DV (2022) Valorization of agricultural wastes for biofuel applications. Heliyon 8(10):e11117

Barisano D, Sharma V K, Braccio G, Singh A and Buddhi D (2020) Performance, cost and market prospects of technologies used for the production, transformation and efficient use of biomethanol.Plant Arc, 20:74-89
Bhardwaj N, Kumar B, Agrawal K, Verma P (2020) Bioconversion of rice straw by synergistic effect of in-house produced ligno-hemicellulolytic enzymes for enhanced bioethanol production. Bioresour Technol Rep 10:100352
Bušić A, Marđetko N, Kundas S, Morzak G, Belskaya H, Ivančić Šantek M, Komes D, Novak S, Šantek B (2018) Bioethanol production from renewable raw materials and its separation and purification: a review. Food Technol Biotechnol 56(3):289–311
Chandel AK, Garlapati VK, Singh AK, Antunes FAF, da Silva SS (2018) The path forward for lignocellulose biorefineries: bottlenecks, solutions, and perspective on commercialization. Bioresour Technol 264:370–381
Cheng CL, Che PY, Chen BY, Lee WJ, Lin CY, Chang JS (2012) Biobutanol production from agricultural waste by an acclimated mixed bacterial microflora. Appl Energy 100:3–9
Dar RA, Arora M, Phutela UG (2019) Optimization of cultural factors of newly isolated microalga Spirulina subsalsa and its co-digestion with paddy straw for enhanced biogas production. Bioresour Technol Rep 5:185–198
Dar RA, Gupta RK, Phutela UG (2021) Enhancement of euryhaline Asterarcys quadricellulare biomass production for improving biogas generation through anaerobic co-digestion with carbon rich substrate. 3 Biotech 11(5):251
Demirbas MF, Balat M, Balat H (2011) Biowastes-to-biofuels. Energy Bioconv Manage 52:15–28
Dong L, Cao G, Zhao L, Liu B, Ren N (2018) Alkali/urea pretreatment of rice straw at low temperature for enhanced biological hydrogen production. Bioresour Technol 267:71–76
Garg V, Dar RA, Phutela UG (2017) Comparative evaluation of different pretreatment methods on biogas production from paddy straw. J Appl Nat Sci 9(3):1525–1533
Gautam P, Upadhyay SN, Dubey SK (2020) Bio-methanol as a renewable fuel from waste biomass: current trends and future perspective. Fuel 273:117783
Gottumukkala LD, Parameswaran B, Valappil SK, Mathiyazhakan K, Pandey A, Sukumaran RK (2013) Biobutanol production from rice straw by a non-acetone producing clostridium sporogenes BE01. Bioresour Technol 145:182–187
Hans M, Kumar S, Chandel AK, Polikarpov I (2019) A review on bioprocessing of paddy straw to ethanol using simultaneous saccharification and fermentation. Process Biochem 85:125–134
Huang Y-F, Lo S-L (2019) Utilization of rice hull and straw. In: Bao J (ed) Rice, 4th edn. AACC International Press, Washington, DC, pp 627–661
Hussy I, Hawkes FR, Dinsdale R, Hawkes DL (2005) Continuous fermentative hydrogen production from sucrose and sugarbeet. Int J Hydrog Energy 30:471–483
Jin X, Song J, Liu GQ (2020) Bioethanol production from rice straw through an enzymatic route mediated by enzymes developed in-house from Aspergillus fumigatus. Energy 190:116395
Karthick C, Nanthagopal K (2021) A comprehensive review on ecological approaches of waste to wealth strategies for production of sustainable biobutanol and its suitability in automotive applications. Energy Convers Manag 239:114219
Kaur K, Phutela UG (2016) Enhancement of paddy straw digestibility and biogas production by sodium hydroxide-microwave pretreatment. Renew Energy 92:178–184
Kaur M, Gill MK, Sharma S, Kocher GS, Sodhi HS (2023) Biological pretreatment strategies for second generation lignocelulosic biomass to enhance ethanol production. In: Srivastava N, Verma B, Mishra P (eds) Agroindustrial waste for green fuel application, Clean energy production technologies. Springer, Singapore, pp 169–203
Keskin A, Emiroglu AO (2010) Catalytic reduction techniques for post-combustion diesel engine exhaust emissions. Energy Educ Sci Technol A 25:87–103
Kim DH, Yoon JJ, Kim SH, Park JH (2022) Acceleration of lactate-utilizing pathway for enhancing biohydrogen production by magnetite supplementation in clostridium butyricum. Bioresour Technol 359:127448

Kiyoshi K, Furukawa M, Seyama T, Kadokura T, Nakazato A, Nakayama S (2015) Butanol production from alkali-pretreated rice straw by co-culture of clostridium thermocellum and clostridium saccharoperbutylacetonicum. Bioresour Technol 186:325–328

Kougias PG, Angelidaki I (2018) Biogas and its opportunities—a review. Front Environ Sci Eng 12:1–12

Kruyeniski J, Ferreira PJ, Carvalho MGVS, Vallejos ME, Felissia FE, Area MC (2019) Physical and chemical characteristics of pretreated slash pine sawdust influence its enzymatic hydrolysis. Ind Crop Prod 130:528–536

Kumar AK, Sharma S (2017) Recent updates on different methods of pretreatment of lignocellulosic feedstocks: a review. Bioresour Bioprocess 4:1–19

Kumari D, Singh R (2018) Pretreatment of lignocellulosic wastes for biofuel production: a critical review. Renew Sust Energ Rev 90:877–891

Liu CM, Wu SY, Chu CY, Chou YP (2014) Biohydrogen production from rice straw hydrolyzate in a continuously external circulating bioreactor. Int J Hydrog Energy 39(33):19317–19322

Liu Y, Yuan Y, Ramya G, Singh SM, Chi NTL, Pugazhendhi A, Xia C, Mathimani T (2022) A review on the promising fuel of the future—biobutanol; the hindrances and future perspectives. Fuel 327:125166

Mechery J, Thomas DM, Kumar CSP et al (2021) Biohydrogen production from acidic and alkaline hydrolysates of paddy straw using locally isolated facultative bacteria through dark fermentation. Biomass Conv Bioref 11:1263–1272. https://doi.org/10.1007/s13399-019-00515-0

Mohapatra S, Mishra RR, Nayak B, Behera BC, Mohapatra PKD (2020) Development of co-culture yeast fermentation for efficient production of biobutanol from rice straw: a useful insight in valorization of agro industrial residues. Bioresour Technol 318:124070

Nair S A and Sivakumar N (2020) Recent advancements in pretreatment technologies of biomass to produce energy. Recent developments in Bioenergy research , 311-324

Nazar M, Xu L, Ullah MW, Moradian JM, Wang Y, Sethupathy S, Iqbal B, Nawaz MZ, Zhu D (2022) Biological delignification of rice straw using laccase from bacillus ligniniphilus L1 for bioethanol production: a clean approach for agro-biomass utilization. J Clean Prod 360:132171

Pal DB, Tiwari AK, Mohammad A, Prasad N, Srivastava N, Srivastava KR, Singh R, Yoon T, Syed A, Bahkali AH, Gupta VK (2022) Enhanced biogas production potential analysis of rice straw: biomass characterization, kinetics and anaerobic co-digestion investigations. Bioresour Technol 358:127391

Pham CT, Boongla Y, Nghiem TD, Le HT, Tang N, Toriba A, Hayakawa K (2019) Emission characteristics of polycyclic aromatic hydrocarbons and nitro-polycyclic aromatic hydrocarbons from open burning of rice straw in the north of Vietnam. Int J Environ Res Public Health 16(13): 2343

Phillips E (2021) Biohydrogen production from cellulosic waste biomass. In: Srivastava N, Srivastava M (eds) Bioenergy research: evaluating strategies for commercialization and sustainability. Wiley, Hoboken, NJ, pp 47–68

Phutela UG, Sahni N, Sooch SS (2011) Fungal degradation of paddy straw for enhancing biogas production. Indian J Sci Technol 4(6):660–665

Ragauskas AJ, Beckham GT, Biddy MJ, Chandra R, Chen F, Davis MF, Davison BH, Dixon RA, Gilna P, Keller M, Langan P, Naskar AK, Saddler JN, Tschaplinski TJ, Tuskan GA, Wyman CE (2014) Lignin valorization: improving lignin processing in the biorefinery. Science 344: 1246843

Rahnama N, Foo HL, Abdul Rahman NA et al (2014) Saccharification of rice straw by cellulase from a local trichoderma harzianum SNRS3 for biobutanol production. BMC Biotechnol 14: 103. https://doi.org/10.1186/s12896-014-0103-y

Robak K, Balcerek M (2018) Review of second generation bioethanol production from residual biomass. Food Technol Biotechnol 56:174–187

Safari M, Tondro H, Zilouei H (2022) Biohydrogen production from diluted-acid hydrolysate of rice straw in a continuous anaerobic packed bed biofilm reactor. Int J Hydrog Energy 47(9): 5879–5890

Sahni N, Phutella UG (2017) Biogas production from paddy straw using thermophilic fungi. Indian J Ecol 44(3):67–70
Sarangi PK, Nanda S, Vo DVN (2020) Technological advancements in the production and application of biomethanol. In: Nanda S, Vo D-VN, Sarangi PK (eds) Biorefinery of alternative resources: targeting green fuels and platform chemicals. Springer, Singapore, pp 127–139
Satlewal A, Agrawal R, Bhagia S, Das P, Ragauskas AJ (2017) Rice straw as a feedstock for biofuels: availability, recalcitrance, and chemical properties. Biofuels Bioprod Bioref 12(1): 83–107. https://doi.org/10.1002/bbb.1818
Sen B, Chou YP, Wu SY, Liu CM (2016) Pretreatment conditions of rice straw for simultaneous hydrogen and ethanol fermentation by mixed culture. Int J Hydrog Energy 41(7):21–28
Sharma A, Singh G, Arya SK (2020) Biofuel from rice straw. J Clean Prod 277:124101
Sheng T, Zhao L, Gao L, Liu W, Wu G, Wu J, Wang A (2018) Enhanced biohydrogen production from nutrient-free anaerobic fermentation medium with edible fungal pretreated rice straw. RSC Adv 8(41):22924–22930
Sindhu R, Binod P, Pandey A, Ankaram S, Duan Y, Awasthi MK (2019) Biofuel production from biomass: toward sustainable development. In: Kumar S, Kumar R, Pandey P (eds) Current developments in biotechnology and bioengineering. Elsevier, Amsterdam, pp 79–92
Singh LK, Chaudhary G (2017) Advances in Biofeedstocks and biofuels. In: Chaudhary G, Singh LK (eds) Biofeedstocks and their processing, vol 1. Wiley Scrivener, Beverly, MA, pp 1–6
Singh A, Sevda S, Abu Reesh IM, Vanbroekhoven K, Rathore D, Pant D (2015) Biohydrogen production from lignocellulosic biomass: technology and sustainability. Energies 8(11): 13062–13080
Srivastava AK, Agrawal P, Rahiman A (2014) Delignification of rice husk and production of bioethanol. Int J Innov Res Sci Eng Technol 3(3):10187–10194
Takano M, Hoshino K (2018) Bioethanol production from rice straw by simultaneous saccharification and fermentation with statistical optimized cellulase cocktail and fermenting fungus. Bioresour Bioprocess 5:16. https://doi.org/10.1186/s40643-018-0203-y
Tian SQ, Zhao RY, Chen ZC (2018) Review of the pretreatment and bioconversion of lignocellulosic biomass from wheat straw materials. Renew Sust Energ Rev 91:483–489
Valles A, Álvarez-Hornos FJ, Martínez-Soria V, Marzal P, Gabaldón C (2020) Comparison of simultaneous saccharification and fermentation and separate hydrolysis and fermentation processes for butanol production from rice straw. Fuel 282:118831
Vamvuka D (2011) Bio-oil, solid and gaseous biofuels from biomass pyrolysis processed an overview. Int J Energy Res 35:835–862
Wang W, Wu X, Chen A, Xie X, Wang Y, Yin C (2016) Mitigating effects of ex situ application of rice straw on CH4 and N2O emissions from paddy-upland coexisting system. Sci Rep 6:37402
Wati L, Narula A (2017) Evaluation of solid state janta biogas plant with paddy straw as supplemental feed. Haryana Agric Univ J Res 47:25–28
Xiao J, Shen L, Zhang Y, Gu J (2009) Integrated analysis of energy, economic, and environmental performance of biomethanol from rice straw in China. Ind Eng Chem Res 48(22):9999–10007
Yoo CG, Yang Y, Pu Y, Meng X, Muchero W, Yee KL, Thompson OA, Rodriguez M, Bali G, Engle NL, Lindquist E, Singan V, Schmutz J, DiFazio SP, Tschaplinski TJ, Tuskan GA, Chen JG, Davison B, Ragauskas AJ (2017) Insights of biomass recalcitrance in natural populus trichocarpa variants for biomass conversion. Green Chem 19:5467–5478
Zeng Y, Zhao S, Yang S, Ding SY (2014) Lignin plays a negative role in the biochemical process for producing lignocellulosic biofuels. Curr Opin Biotechnol 27:38–45
Zoghlami A, Paes G (2019) Lignocellulosic biomass: understanding recalcitrance and predicting hydrolysis. Front Chem 874:1–11

Chapter 5
Lignocellulosic Waste to Biofuel-Paddy Straw to Bioethanol: Advancement in Technology

Irum, Mohommad Asif, Priyanka Dubey, Gyanendra Tripathi, Nortoji A. Khujamshukurov, Alvina Farooqui, and Vishal Mishra

Abstract Paddy is amongst the most consumable crops in the world. In fact, in Asia, it is known as the most stable food item. As per the growing population, there is a need for the large amount of production of rice crops which also generates large amounts of rice straw along with the rice as residue in the fields. Nowadays, farmers and others are using this rice straw unsustainably by simply burning these paddy straws in the fields which directly contributes to the population as it emits large amounts of greenhouse gas in the environment. So, there is a need for the sustainable development of these paddy straws to convert them into good end products to make them used appropriately. Utilizing paddy straw as a valuable source of lignocellulosic biomass presents a promising method for generating bioethanol. This sustainable use of paddy straw for bioethanol will also bring lots of additional income from the waste to the country. The present chapter comprises components of lignocellulose biomass (LCB) in rice straw and different steps to produce bioethanol.

Keywords Paddy straw · Lignocellulose biomass · Bioethanol · Pretreatment · Fermentation

Irum
Department of Biosciences, Integral University, Lucknow, Uttar Pradesh, India

M. Asif · P. Dubey · G. Tripathi · A. Farooqui (✉)
Department of Bioengineering, Integral University, Lucknow, Uttar Pradesh, India
e-mail: alvina@iul.ac.in

N. A. Khujamshukurov
Tashkent Institute of Chemical Technology, Tashkent, Republic of Uzbekistan

V. Mishra (✉)
School of Biochemical Engineering, IIT (BHU), Varanasi, Uttar Pradesh, India
e-mail: Vishal.bce@itbhu.ac.in

N. Srivastava et al. (eds.), *Paddy Straw Waste for Biorefinery Applications*, Clean Energy Production Technologies, https://doi.org/10.1007/978-981-99-8224-0_5

5.1 Introduction

The dependency of the world upon fossil fuels is growing gradually nowadays, to fulfil its need for energy, and about 80% of worldwide energy is generated by burning these fossil fuels; amongst them 58% of energy is expended in the carriage section (Tripathi et al. 2022). Countries are facing problems regarding the fuel availability fuel. Primarily, the hasty consumption of fossil fuels in various sectors is directly connected to its fast exhaustion (Agarwal 2007). The depletion of fossil fuel reserves on Earth is anticipated to occur within the next 40 to 50 years (Zabed et al. 2016). Secondly, greenhouse emissions and global warming are directly connected to fossil fuel combustion which is directly related to environmental changes such as sea level rise, biodiversity rise, pollution, and others (Singh et al. 2010). Last but not least, the political crisis also led us to rethink our dependence on fossil fuels; meanwhile these crises resulted in oil supply disruption, thus troubling energy sectors of both developing and developed countries (Mussatto et al. 2010). So, it is getting important to find a compatible alternative source for various sectors of energy consumption. Lignocellulosic biomass (LCB) is renewable, sustainable, effective, and cheap with lesser greenhouse emissions and a better way of producing biofuel. LCB has become an alternative to fossil fuel due to its affluence of producing biofuel with less toxicity (Zabed et al. 2014; Dubey and Yousuf 2021). The conversion efficiency and bioethanol production yield are dependent on the nature and content of LCB (Tye et al. 2016).

Plant biomass is chiefly composed of lignocellulose and is plentiful from various renewable feedstocks such as agricultural waste, forestry wastes, municipal solid waste, and dedicated energy crops (Singh et al. 2022). Globally, LCB is up to 10 to 50×10^9 t/years (Sticklen 2006). Paddy straw availability is about 731.3 Tg/annum and 204.6 GL bioethanol (Volynets et al. 2017).

When accounting for both unused crops and entire residual crops, the annual production of bioethanol from lignocellulosic biomass is projected to reach around 450 billion litres approximately (Kim and Dale 2004). The USA individually produces 1368 million tons of biomass for the bioethanol production from agricultural waste that is about 428 MT, whereas forest waste contributes 370 MT, energy crops 377 MT, 87 MT from grains and crops, 58 MT from municipal and industrial wastes, and other wastes 48 MT (Saini et al. 2015).

To put it more precisely, the energy consumption is estimated to rise by 49% globally, between 2007 and 2035. This increase is attributed to the simultaneous growth of the economy, expansion of the population, and societal demands (Prasad et al. 2016). A significant portion of this substantial rise can be attributed to India and China (Bhardwaj et al. 2020). Bioethanol has considerable oxygen content, leading to improved combustion efficiency. Furthermore, its heightened octane rating facilitates the efficient operation of engines through increased compression ratios (Branco et al. 2019). Bioethanol can be derived from various sources like maize, sugarcane, grains, and sugar beets. The global bioethanol production is primarily divided between Brazil and the USA, accounting for approximately 90% of the total.

Brazil contributes to over 50%, while the USA's share is under 30% (Branco et al. 2019). The complex constituents of lignocellulosic material comprise three core constituents: lignin, cellulose, and hemicellulose. Cellulose, which consists of glucose polymers, provides plants with structural stability. Hemicellulose is responsible for bonding, while lignin contributes to the overall strength of the structure (Bhardwaj et al. 2020). Consequently, these feedstocks necessitate labour-intensive and expensive pretreatment processes before undergoing hydrolysis, fermentation, and distillation. Pretreatment indeed holds a crucial role in this regard as it enhances the extent of the non-crystalline area to facilitate the hydrolysis process and increase the porous matrix's porosity to enable the processes of chemical and enzymatic hydrolysis to effectively break the bonds between cellulose and lignin and hemicellulose (Tran et al. 2019). Within this chapter, we will explore various parameters involved in the bioethanol production process.

5.2 Lignocellulosic Biomass

Lignocellulosic biomasses (LCB) are a sustainable resource of renewable bioenergy and are produced in huge quantities mainly by plants. They comprise forestry or wood waste and agricultural waste which majorly contain cellulose, lignin, tannin, saponin, and hemicellulose. Cellulose, constituting a significant portion ranging from 30% to 50% of the biomass's total dry weight, is subjected to degradation by the enzyme cellulase. This degradation leads to the conversion of cellulose into glucose, eventually transformed into biofuel (Harris and DeBolt 2010). Aside from cellulose, hemicelluloses exhibit an amorphous character attributed to the presence of heteropolymers like glucomannan, xylan, and arabinoxylan (Cywar et al. 2022). Hemicelluloses can be hydrolysed using diverse methods such as acid hydrolysis, base hydrolysis, and enzymatic hydrolysis. Within LCB, lignin stands as the second most abundant carbon source, following cellulose.

Globally, most common agricultural wastes are collected from wheat, rice, sugarcane, etc., whereas the rest of the agricultural waste contributes to a very low volume of worldwide biomass. The prevalent leftover material for bioethanol production is corn stover, which amounts to around 1.0 kg/kg of corn grain or 4.0 t/acre. This residual by-product stems from the corn kernel and encompasses components such as stalks, leaves, cobs, and husks (Koul et al. 2022). Wheat straw yield is about 1–3 t/annum/acre. On the other hand, rice straw is regarded as an exceptionally abundant form of lignocellulosic biomass, with a global estimated quantity of 731 million metric tons per year (Saini et al. 2015). Bagasse is obtained during the processing of sugarcane which is also an important agro-waste with an estimated global yield of 317–380 MT/year (Sánchez 2009). In Asia, mostly rice and wheat straw are produced, while America produces large amounts of bagasse and corn stover.

Several energy crops are used for the production of bioethanol. Amongst them, various C_3 (*Miscanthus* sp., *Panicum virgatum*, etc.) and C_4 (*Arundo donax*,

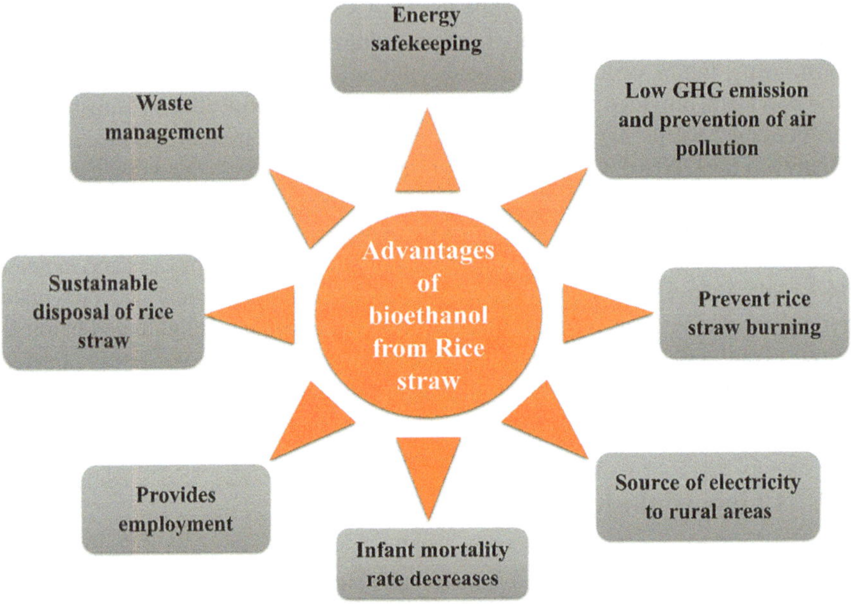

Fig. 5.1 Advantages of bioethanol from rice straw

Medicago sativa, etc.) plants are there (Lewandowski et al. 2003). Forest biomass is mainly composed of woody substances, while forest wastes comprise dead tree branches, sawdust, wood chips, barks, etc. (Limayem and Ricke 2012). In the USA, approximately 370 MT/year of forest woody feedstock is produced (Perlack et al. 2005). Countries like Canada, Russia, Brazil, and China collectively contribute to half of the whole forest area worldwide (Hadar 2013). Municipal solid wastes also work as a good source for biofuel production. In 1990, it was reported that each country approximately produces 250 kg of this waste which has a total global production of approximately 1.3×10^9 t, and this data should be doubled in the next 10 years (Hadar 2013). However, these are not ideal waste materials due to their variety of composition and heavy contamination chances (Schmitt et al. 2012). Figure 5.1 shows the advantages of the bioethanol produced from the rice straw. Although there are great advantages of bioethanol from rice straw, it also has some cons including the elevated transportation expenses and the utilization of gasoline/diesel for conveyance and various applications, significant expenses associated with pretreatment, the release of nitrous oxide resulting from the breakdown of nitrogen-rich fertilizers, and emission of methane from paddy fields (Singh et al. 2016). Figure 5.1 represents the advantages of bioethanol produced from rice straw.

5.2.1 *Components of LCB*

Lignocellulosic biomasses mainly comprise holocellulose (celluloses and hemicelluloses) and lignins from which holocellulose comprises one third of the total LCB dry weight and is also a chief material utilized as a foundation for generating bioethanol (Cavka and Jönsson 2013). The content of LCB varies depending on its source and physical properties of biomass. Holocellulose, i.e. cellulose and hemicellulose, are bonded with lignin molecules via covalent interaction and H-bonding that enables the whole arrangement vigorous and also intractable, and the recalcitrance of LCB is due to the crystalline cellulose, available surface area, lignin protection, and particle's heterogeneous characters (Mosier et al. 2005; Bhatia et al. 2020). Some additional components (pectin, polyamino acid chain, extractives, ash, etc.) are also found, but it is in very less amount (Menon and Rao 2012; Schutyser et al. 2018), and they are found within the cell wall's middle lamella. Generally, cell walls have different layers and are classified as primary walls and secondary walls. The primary wall is very thin and has fewer amounts of cellulose and lignin but a very large amount of pectin and hemicellulose. While the secondary wall which is also subdivided is rich in cellulose, middle lamella has an abundance of lignin (Pandey 2008; Aitken 2012; Gibson 2012). Table 5.1 represents the percentage of the lignocellulosic components in different feedstocks.

5.2.1.1 Cellulose

Linear D-glucose homopolymer is associated through glycosidic linkages to form a chain, and these chains vary from 100 to 10,000 monomers (Zhao et al. 2012) and have an average weight of 100,000 (Saxena et al. 2009). Cellulose is water-insoluble and doesn't undergo depolymerization due to the presence of hydrogen bonds which results in a crystalline matrix structure and microfibrils which are formed by the cross-linking of hydroxyl groups. The primary cell wall has 15–30% dry weight of cellulose, while the secondary cell wall has 40% of dry weight (Zhao et al. 2012). While other starchy materials need high temperatures to become amorphous,

Table 5.1 Lignocellulosic components in different feedstocks

S. no.	Lignocellulose waste	Cellulose (%)	Hemicellulose (%)	Lignin (%)	References
1.	Rice straw	31.6	29.2	9.8	Watanabe et al. (1993)
2.	Barley husk	>20	>35	<20	Parajó et al. (2004)
3.	Wheat straw	48.6	27.7	8.2	Saha et al. (2005)
4.	Sunflower stalk	>40	<30	>13	Jiménez et al. (1990)
5.	Sorghum straw	35	>20	25	Vazquez et al. (2007)
6.	Tall fescue straw	31	20	14	Kumar and Murthy (2011)
7.	Barley straw	>30	>25	>15	Nabarlatz et al. (2007)

cellulose only requires 32 °C and 25 MPa pressure to be converted as an amorphous molecule (Ashokkumar et al. 2022).

5.2.1.2 Hemicellulose

It is a short, linear, and branched heteropolymer of simple sugars. Major types of monomers are hexoses, pentoses, and some sugar acids that are present and build up the structure of hemicelluloses (Ravindran and Jaiswal 2016). Xylan is the backbone of hemicellulose in which D–xylose and L–arabinose are linked together through β-1, 4 linkages (Mood et al. 2013). Xylan and glucomannans are the most commonly found hemicelluloses in the biomass. Hemicellulose acts as a bonding agent between celluloses and lignin via non-covalent and covalent interaction, respectively. In this manner, it helps to maintain the integrity of the biomass and also protects the fibres from cellulase enzymes (Michelin et al. 2013).

5.2.1.3 Lignin

It is an extremely branched aromatic polymer especially found amongst woody biomass (Drummond and Drummond 1996); mostly it is made up of three phenolic compounds. Lignin is amorphous in nature as compared to the other contents of biomass (Aitken 2012) with a molecular weight of about 10,000 Da (Mielenz 2001). Woody biomass is rich in cellulose and lignin amongst which softwood barks have high lignin content in comparison to hardwood, whereas grasses, agricultural residues, etc. comprise the lowest lignin content. A higher amount of lignin imparts strengths plus rigidity to plants which also provide impermeability and work as a physicochemical barrier against microbes (van Dyk and Pletschke 2012).

5.2.1.4 Paddy Straw as LCB

Paddy straw is a by-product of rice cultivation and represents an abundant bio-resource. Paddy straw is amongst the richest LCB waste material on the Earth. In Egypt, it delivered around 4 MT. Paddy straws might create 205 billion litres of bioethanol each year on this planet, which is around 5% of all out of utilization (Ballerini et al. 1994). It is the biggest sum from a solitary biomass feedstock, as indicated by a new report led by the Public Organization for Interdisciplinary Science and Innovation (NIIST), India. One of the most plentiful inexhaustible assets for bioethanol creation is paddy straw with high sugar content. Regardless of this, the executive of this humungous measure of paddy straw is the reason for the main issue as a legitimate foundation for capacity objects is inaccessible. Subsequently, a huge piece of this straw is singed in open fields, which floods the discharges of ozone-harming substances, yet, in addition, influences soil wellbeing. One of the most plentiful inexhaustible assets for bioethanol creation is paddy straw with high sugar content. Regardless of this, the executive of this humungous measure of paddy straw is the reason for the main issue as a legitimate foundation for capacity objects is inaccessible. Subsequently, a huge piece of this straw is singed in open fields, which floods the discharges of ozone-harming substances, yet, in addition, influences soil wellbeing (Gadde et al. 2009a, 2009b).

Oryza sativa (rice) is one of Asia's staple food. According to a study, in 2011 annual rice production in Asian countries was about 721.4 MT and 90.48% (Khan 2012) which produced 973.89 MT of rice straw (Kadam et al. 2000). Even though, amongst the cereal crops, rice comes in second position and wheat remains the first, still rice produces a greater amount of residues approximately 330 MMT (Robinson 2006). Amongst all other LCB wastes, cereal straws are utmost ample, cost-effective, renewable, and easy to obtain (Park et al. 2011). On the planet, amongst the agricultural harvest deposits, paddy straw becomes the most accessible cellulosic source. A chief underlying constituent of straw's cell wall is microfibrils. The cellulosic chains of the cell wall are consolidated with H-bonds to form microfibrils of crystalline nature. Microfibrils give mechanical solidarity to cell walls and are reinforced in a matrix made up of hemicelluloses, lignins, and carbohydrates (Atanu et al. 2006). The structure of paddy straw is made up of celluloses, hemicellulose, lignins, ash, and silica, while the amount of the content depends on the environment where it grows and the type of variety used. Generally, 33–47% of cellulose, approximately 30% of hemicellulose, 5–25% lignin (Saha 2003), and up to 18.3% of silica are present in the paddy straw (Hessien et al. 2009).

The pentose sugars prevail in hemicellulose which mainly contains xylose. This is the main sugar monitored by arabinose and hexoses. The starch content in paddy straw includes the desired amount of various carbohydrates (Binod et al. 2010). One of the most plentiful inexhaustible assets for bioethanol creation is paddy straw with high sugar content. Regardless of this, the executive of this humungous measure of paddy straw is the reason for the main issue as a legitimate foundation for capacity objects is inaccessible. Subsequently, a huge piece of this straw is singed in open fields, which floods the discharges of ozone-harming substances, yet, in addition, influences soil wellbeing (Gadde et al. 2009a, 2009b).

5.3 Stages in Bioethanol Production

The procedure of bioethanol manufacturing from various sources includes pretreatment, hydrolysis, fermentation, and ethanol recovery (Fig. 5.2). These processes are mentioned below.

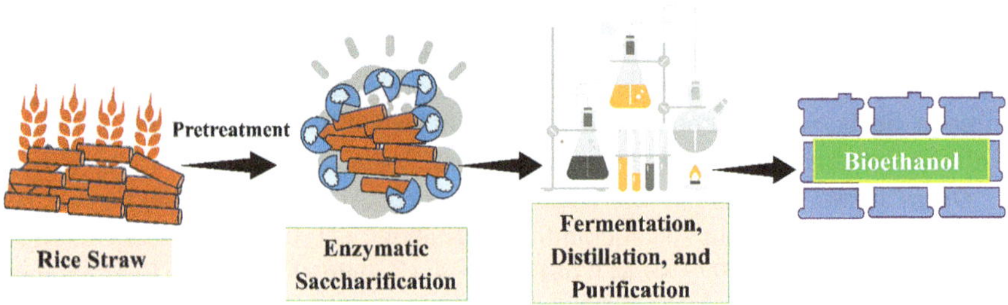

Fig. 5.2 Steps for the production of bioethanol

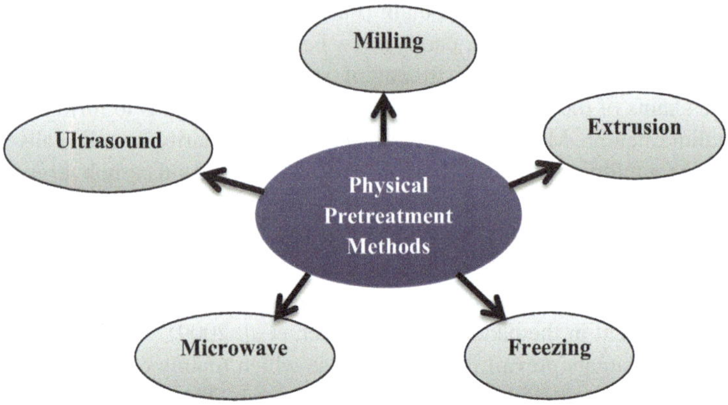

Fig. 5.3 Different physical pretreatment methods involved in the production of bioethanol

5.3.1 Pretreatment

A group of methods has been proposed during the previous decades and is broadly categorized into many groups like physical, chemical, physiochemical, and biological or combination type of pretreatment (Farooqui et al. 2021).

5.3.1.1 Physical Pretreatment Methods

The type of physical pretreatment (i.e. chopping, milling, etc.) aims for the reduction of the particle size of the LCB proportionally to raise the surface area of the biomass which facilitates enzymatic hydrolysis. Various types of physical pretreatment methods are given below in Fig. 5.3. The power required to acquire maximum simplification in enzymatic hydrolysis is typically probability high. Hypothetical energy content that is present in the biomass can be even much higher (Hsu 2018). However, combining physical pretreatment with another type of pretreatment might be an interesting choice (Galbe and Zacchi 2007). γ-Rays are another technique that has been suggested, and it involves irradiating cellulose that breaks glycosidic bonds and, thus, lowers crystal nature and provides maximum surface (Takacs et al. 2000). To use this method in a large-scale process, however, would be extremely expensive. Additionally, it is questionable whether it can be used in conjunction with environmentally friendly technologies. Figure 5.3 shows different physical pretreatment methods involved in the production of bioethanol.

Electron Beam Radiation

Cellulose in LCB gets despoiled into several fragile fibres, its constituent oligomeric sugar, cellobioses, etc. when it undergoes irradiation (Kumakura and Kaetsu 1983). This is due to the breaking of glycosidic linkage by irradiation on cellulose molecules. This physical method of pretreatment is expensive and consumes high energy. Bak (2014) concluded that pretreatment on dry-milled rice straw utilizes e^{-}beam treatment. On doing enzyme hydrolysis of previously irradiated rice straw, the yield of reducing glucose increases several times while comparing with untreated rice straw after 5 days and 5 h hydrolysis. Physical changes are shown in the treated rice straw while performing this type of pretreatment under SEM and X-ray diffraction analysis.

Grinding and Milling

This type of pretreatment is done to minimize the particle size of the desired biomass. The blending of this physical pretreatment method along with other types of pretreatment has been done by which the crystal nature of the biomass gets interrupted. According to a study conducted by Jin and Chen in (2006), they found that during the hydrolysis process, using superfine ground biomass was more effective than using biomass that had been ground normally. In other words, finely grinding the biomass to a very small particle size yielded better results in terms of the hydrolysis process compared to using biomass with a coarser grind. However, the energy essential for this procedure is also to be considered while making it commercialized. Wet disk milling in the case of rice straw pretreatment is much better than that of ball milling in respect of energy consumption and glucose reclamation (Hideno et al. 2009). There are many more types of grinding and milling, but lack of reports is there with regard to rice straw.

Microwave Pretreatment

This physical pretreatment method has high efficiency and can be easily operated. With this technique, changes in the cellulose content (Xiong et al. 2000) and degradation of hemicellulose and lignin occur, and enzymatic susceptibility gets increased (Azuma et al. 1984). Microwave pretreatment could enhance the enzymatic hydrolysis of rice straw in water as a medium (Rezania et al. 2020), likewise in a medium rich in glycerine.

5.3.1.2 Chemical Pretreatment Methods

Branched and unbranched cellulosic bonds create robust hurdles that prevent enzymes from entering the fibres. The basic pretreatment (alkaline) is done by emerging the biomass in an alkaline medium (like NaOH) through which swelling of the pore can be achieved and then placing it at a high temperature for a definite time interval. Internal surface area is increased due to the swelling caused by the alkaline solution and a reduction in the amount of crystallinity and polymerization. Generally, a key part of lignin along with some hemicellulose is being solubilized together. However, a huge amount of hemicelluloses are obtained as oligomers. Pretreatment with an alkaline medium interrupts lignin assembly by breaking bonds between lignin and the carbohydrates, which makes the availability of carbohydrates for enzymatic pretreatment. As it acts as a removal of lignin content, when compared to wood products, it works better on agricultural waste and herbaceous crops, as we know these materials commonly retain less lignin. For softwood species with high lignin content, there is an observation of very little or negligible effect. However, using lime for pretreatment is an additional alkaline method, mainly fitting for agricultural wastes like hardwood materials, corn stover, etc. (Kim and Holtzapple 2005). A variety of chemical pretreatment methods are given in Fig. 5.4. By immersing the biomass in an acidic solution then after heating it at 140–200 °C for desired amount of time, diluted acidic pretreatment is carried out with sulphuric acid concentration below 4 wt.%. Due to its low cost and effectiveness, it has attracted the most interest in these studies. Mainly monomer sugars are obtained by the hydrolysis of hemicelluloses. It was observed that materials that undergo acid hydrolysis are thought to get fermented because of the existence of poisonous compounds (Bajpai and Bajpai 2016).

Another way of chemical pretreatment is the use of an organic solvent mixture in the presence of organic compounds (HCl/H_2SO_4), which are intended to lyse the bonds of hemicellulose or lignin. This procedure is generally stated as the

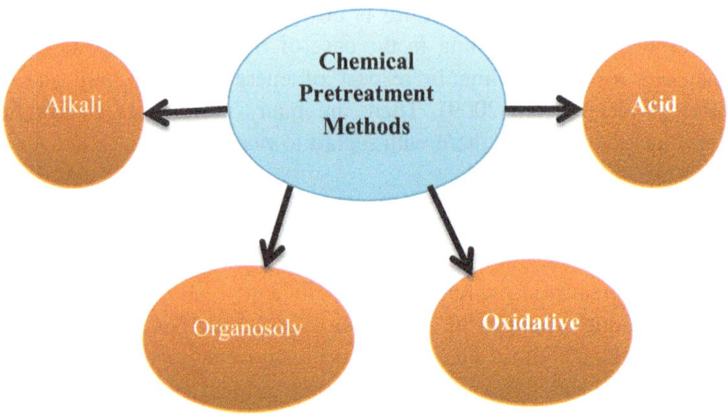

Fig. 5.4 Different chemical pretreatment methods used in bioethanol production

organosolvent method (Pan et al. 2005). These situations involve the dissolution and recovery of the hydrolysed lignin during the organophilic phase. Before undergoing the further downstream process of the bioethanol production process, the material must be cleaned as the solvents could work in inhibitors aspects. Solvents like ethanol, methanol, acetone, ethylene glycol, phenol, etc. are used. These solvents are difficult to handle because some are explosive and highly inflammable. There are many ways of chemical pretreatment amongst which some are shown in the Fig. 5.4.

5.3.1.3 Physicochemical Methods of Pretreatment

This method of classification involves a mixture of chemical and physical pretreatment techniques. For the lignocellulosic biomass, pretreatment with steam method is broadly used. It is also called an explosive method because it was believed that to make the material viable for hydrolysis, the explosive action on the fibres of the materials is needed. It was proved that the steam pretreatment effect occurs by hemicellulose's hydrolysis in the presence of acids, which elucidates how cellulose gets easily decomposed in comparison to others (Knappert et al. 1980; Brownell et al. 1986). Particularly, organic acids found in some types of hardwood and agricultural wastes serve as catalysts for the hydrolysis of hemicellulose. Material is put under high-pressure steam between approximately 150 and 250 °C and 6–34 bar pressure by using steam pretreatment for the specific period, and then pressure is released. Mainly hemicellulosic content of the LCB gets dissolved during this pretreatment as well as can be procured from the liquid phase in the form of oligomers or monomers. Cellulosic content becomes more accessible to the enzymes when present in the solid phase. In some cases, it can be challenging to discover an environment that produces the maximum amount of both C_6 and C_5 sugars while also producing a cellulosic content that is simple to hydrolyse into hexose sugar (glucose). Thus, a two-step steam pretreatment process is needed where the cellulose fraction is pretreated more severely, while the hemicellulose sugars are recovered with less severity. Figure 5.5 represents various types of physicochemical pretreatment methods.

H_2SO_4 or SO_2 (acid catalysts) improves this pretreatment method (steam). These acids aid in improving enzyme hydrolysis as well as hemicellulose recovery. Since softwood generally has a harder time degrading than other types of wood, it is crucial to use an acid catalyst when working with softwood. Thus, pretreatment for enhancing hydrolysis and enzymatic digestibility of the raw materials that are most likely to be commercialization is steam pretreatment in the presence of a catalyst, and it has also undergone extensive testing using pilot-scale equipment (Galbe and Zacchi 2007). Hydrothermolysis is referred to as water treatment at high degrees. It is known as LHW (liquid hot water) pretreatment and involves high-temperature water treatment. The process is somehow comparable to steam pretreatment; however, it uses a minor temperature and fewer dry residues, thus recovering more polysaccharides and oligosaccharides as a result (Bouchard et al. 1991; Griebl et al. 2005). Modification in this method is done by the addition of an acid which makes it

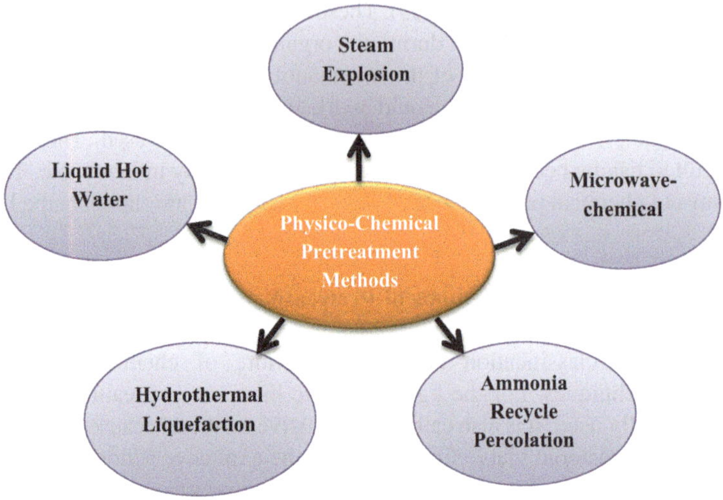

Fig. 5.5 Different physicochemical pretreatment methods

similar to diluted acid pretreatment. Because there is significantly higher water content than during steam pretreatment, the saccharide solution thus obtained is more diluted which makes the subsequent process highly energy-intensive. Processing of bagasse at a high temperature (220 °C) and then undergoing fermentation using baker's yeast (*S. cerevisiae*), there was essentially no variation in ethanol production in the range of 1–10 wt.% dry matter (Mood et al. 2013). With wet oxidation pretreatment, the biomass gets heated at a temperature above 120 °C while being treated with H_2O or O_2 and occasionally an alkali catalyst. Due to the yield being lowered with increasing amounts of lignin and the fact that a significant portion of lignin gets dissolved, this is a best practice appropriate for low lignin-content materials (Schmidt and Thomsen 1998). This inability of lignin to work like solid fuel with many other delignification processes significantly lowers the revenue on industrial-level end products.

Parallel to this steam pretreatment method, the AFEX (alkaline ammonia fibre explosion) method also works at high pressure. The dry matter is added to liquid ammonia for 10–60 min at below 100 °C and more than 3 MPa (Teymouri et al. 2005). Up to 2 kg/kg ammonia and dry biomass are used. Since ammonia is amongst the highly volatile chemicals at atmospheric pressure, it can easily be recycled just by dropping the pressure. Only a small portion of the dry matter is dissolved during pretreatment, leaving approximately no hemicellulose and lignin behind. The hemicellulose is degraded into simpler sugars and other modifications (Gollapalli et al. 2002), which makes it insoluble. However, the material's structure is altered, leading to a higher water-holding capacity and digestibility. AFEX works best on agricultural waste, like the other alkaline pretreatment techniques, but does not affect wood due to its higher amount of lignin in it (Wyman 1996; Belkacemi et al. 1998). According to a study of this process, no inhibitor formation occurs which could alter the downstream (Sun and Cheng 2002). APR (ammonia recycle percolation) technique is a different kind of ammonia-using process (Iyer et al. 1996). During this procedure, a solution of aqueous ammonia (with a concentration of 10–15 wt.%) is circulated through the biomass under elevated temperatures (ranging from 150 to 170 °C), following which the ammonia is reclaimed. ARP stands as an effective delignification technique for hardwood and agricultural residues, yet it demonstrates relatively lower efficacy with softwood. Figure 5.5 represents different physicochemical pretreatment methods.

5.3.1.4 Biological Method of Pretreatment

Application of lignin-digesting microorganisms, like fungi and some bacteria, can be used as biological agents in biological pretreatment (Lee 1997). Thus this method is environmentally friendly and energy efficient since it is achieved at minimum temperature lacking any chemical use. It is a slow process as per its utilization in industries. Meanwhile, these microorganisms consume the components of the LCB (hemicellulose, cellulose, lignin) to various extents, and there is a gentle amount of material loss (Wyman 1996). However, the technique could be applied as a first step before utilizing some of the other pretreatment techniques.

5.3.1.5 Detoxification

While the pretreatment phase plays an essential role in the conversion of lignocellulosic biomass (LCB) into ethanol, it brings along a significant drawback to the overall process. This drawback arises from the fact that the conditions of pretreatment give rise to the formation of by-products originating from lignocellulose. These by-products can pose inhibition challenges for both fermenting microorganisms and enzymes between the process steps, particularly when their accumulation reaches noteworthy levels (Bhatia et al. 2020). The pretreatment procedure leads to the creation of diverse by-products, including sugar acids, acetic acid, formic acid, hydroxymethylfurfural (HMF), furfural, and levulinic acid (Jönsson et al. 2013). The majority of inhibitors that emerge from lignocellulose trace their origins to the degradation of hemicelluloses and lignin during the pretreatment phase. Moreover, the presence of cellulose and biomass extractives can inadvertently contribute to the formation of inhibitors due to the effects of the pretreatment conditions. These inhibitors stemming from the pretreatment stage can be categorized into three primary groups based on their source: aliphatic acids, furan derivatives, and phenolic compounds (Bhatia et al. 2020). A variety of approaches have been put forth to tackle the challenges posed by inhibitors. These strategies encompass alternative procedures for the conversion process as well as treatments applied to the pretreated biomass with the aim of mitigating or neutralizing the impact of inhibitors. These treatments encompass a range of biological, physical, and chemical techniques. However, the specific types and quantities of inhibitors are influenced by the specific pretreatment methods employed. This consideration is carefully weighed before opting for a detoxification strategy.

Detoxification becomes imperative when faced with scenarios involving profoundly inhibitory hydrolysates during fermentation. This situation arises when a significant build-up of inhibitors occurs within the fermentation mixture when a fermenting microorganism with restricted tolerance to inhibitor is employed. The detoxification procedure needs to be strategically devised to specifically target the removal of inhibitors. Moreover, it should be economically viable and seamlessly integrable into the comprehensive process (Kumar et al. 2020).

5.3.2 *Enzymatic Saccharification and Fermentation*

Once the paddy straw (rice) pretreatment is done, a hydrolysis (or enzymatic saccharification) process is initiated which uses enzymes to split cellulose and hemicellulose polymers. After that fermentation occurs, this transforms the sugars into ethanol. The amount of sugar recovered determines how much ethanol is produced, the kind of modest sugar (pentose or hexose), and inhibitor formation (Singh et al. 2014a). Glucose is produced by the hydrolysis of cellulose, while several pentose and hexose are released by the hemicellulose (Taherzadeh and Niklasson 2004). Enzymatic saccharification is impacted by various factors, encompassing the substrate's composition, the method of pretreatment employed, enzyme dosage, the efficacy of hydrolytic enzymes, and more. The cellulose's degree of polymerization and crystallinity significantly influence the pace at which enzymatic hydrolysis occurs (Chang and Holtzapple 2000). The presence of hemicellulose has an impact on enzymatic hydrolysis because when hemicellulose is removed, it leads to an increase in the average pore size of the substrate. This larger pore size, in turn, speeds up the enzymatic hydrolysis process, making the substrate more accessible to the enzymes, and ultimately, this increased accessibility further accelerates the rate of hydrolysis (Chandra et al. 2007). The maximum yields of approximately 90%, 70%, and 80% for total carbohydrate, xylan, and sugars, respectively, were achieved at a condition of the enzyme (40 FPU/g-substrate), sodium sulphite (12%), and temperature (160 °C). Enzymatic saccharification increased by adding sodium sulfite-formaldehyde to pretreated rice straw (Gu et al. 2013). Research by Yao et al. (2011) on SO_3 collaborative dilutes alkali approach suggested that SO_3 gas initially penetrates the internal structure. Subsequently, it interacts with the moisture within the straw, leading to an internal micro-thermal explosion. After removing the lignin with a weak alkali solution, the rice straw was saccharified at a rate of 91% based on pretreatment with SO_3 for 4 h and subsequent treatment with NaOH (1% w/v) for about 7 h up to 50 °C. By removing lignin and hemicelluloses, rice straw's enzymatic saccharification also increases by microwave-assisted chemical pretreatment (Singh et al. 2013, 2014b). Sulbarán-de-Ferrer et al. (2003) evaluated that samples of rice straw that had been treated with ammonia had a higher percentage of monomeric sugars (61%) than rice straw which is untreated (11%).

The rice straw either undergoes SSF (simultaneously saccharified and fermented) or SHF (separately hydrolysed and fermented) to produce ethanol. Gupte et al. (2022) concluded the SHF processing of rice straw by biological agents. Using a technique where saccharification and fermentation are carried out simultaneously is better than the approach where hydrolysis and fermentation are done separately. This is because the simultaneous method can eliminate the issue of end-product inhibition during the saccharification process and also reduces the overall processing time. As a result, this approach increases the yield of ethanol, making it a more efficient and effective method for ethanol production (Alfani et al. 2000; Sarkar et al. 2012). Various optimum temperatures for hydrolysis and fermentation are a major drawback of the SSF process. The ideal temperatures for hydrolysing enzymes and yeast for fermentation are, respectively, 30–35 and 50 °C (Kádár et al. 2004). The bacterium *Zymomonas mobilis* and the yeast *Saccharomyces cerevisiae* are well-known microorganisms for ethanol production (Claassen et al. 1999) and thus deal with high ethanol production (of about ≥90% theoretically). According to research performed by Jørgensen (2009), it was concluded that the yield of bioethanol via fungus (native) is up to 99%. Ethanol undergoes enzymatic hydrolysis via enzymes, fermentation, distillation, and dehydration to achieve 99% purity. In a boiler, lignin, unreacted saccharides, and some organic materials are burned to create process steam and electricity.

5.3.3 Product Recovery

The recovery of ethanol from the fermentation blend is accomplished through either sole distillation or a combination of distillation and adsorption methods (Rahmati et al. 2020). Distillation serves as a long-standing approach for ethanol retrieval. The initial step involves subjecting the fermentation mixture to distillation, leading to the separation of ethanol from water and ultimately achieving an ethanol with 95% concentration approximately. The resulting liquid at the bottom, containing lignin residues, unreacted hemicellulose and cellulose, enzymes, ash, microorganisms, and other constituents, is subsequently directed to wastewater treatment (Robak and Balcerek 2020). These components have the potential to be concentrated and harnessed as a process-driving fuel source or transformed into diverse by-products.

Bioethanol exhibits favourable attributes, including a higher-octane number, notable heat of vapourization, and a low cetane number, rendering it well-suited for amalgamation with gasoline. This blending approach contributes to reduced emissions of particulate matter and nitrogen oxides, attributed to its 35% oxygen content. Diverse countries have embraced distinct bioethanol-gasoline blending ratios, with E85 (consisting of 85% bioethanol and 15% gasoline) gaining popularity for light-duty vehicles (Mat Aron et al. 2020). In India, the initiation of the ethanol petrol blending policy (EBPP) in 2003 aimed to promote ethanol utilization, evolving over time to mandate blending requirements of up to 20% by 2017. However, achieving these targets solely with ethanol derived from sugarcane molasses poses challenges due to supply constraints. To surmount obstacles such as taxation issues, the National Biofuel Policy introduced tax concessions for bioethanol. The utilization of lignocellulosic biomass, such as rice straw, for bioethanol production emerges as an environmentally conscious and sustainable avenue to meet blending goals (Singh and Basak 2019).

5.4 Conclusion

The production of bioethanol from the lignocellulosic biomass requires an eco-friendly and cheap alternative for its production on a large scale. Taking the essentiality of second-generation biofuels into consideration, rice straw has appeared as the most potent source for bioethanol production, owing to its availability in abundance and attractive composition. Employment of hydrolysing enzymes for the conversion of rice straw in the presence of biological agents into fermentable sugars is the utmost smart practice nowadays from environmental point of view. However, factors like the complex nature of lignocellulosic biomass, high ash content, and high lignin content cause hindrances in the way of developing a cost-effective and feasible technology for biofuel production. Numerous research are going on for the development of efficient pretreatment methods to eradicate the undesirable portion of lignocellulosic biomass. This will help to get the sugars readily available for the bioethanol production. Considerable success has been achieved by various pretreatment methods to date. The available research outcomes have evident the need for bioethanol in transport areas that could be achieved by the utilization of rice straw. More studies in the area of the process as well as strain engineering still need to be explored for the improvement of the system efficacy and evasion of the complications related to the combined fermentation of different types of sugars (xylose, glucose, etc.). A much synchronized combination of all the above steps for the production of bioethanol (pretreatment, hydrolysis, and fermentation) needs to be selected to obtain the high efficiency of the method. The process of bioethanol production using rice straw with the incorporation of advancements such as GMO yeast, synthetic hydrolysing enzymes, and other technologies in the efficient combination has the potential to be the most feasible technology in the future.

References

Agarwal AK (2007) Biofuels (alcohols and biodiesel) applications as fuels for internal combustion engines. Prog Energy Combust Sci 33:233–271

Aitken S (2012) Wood chemistry and secondary cell wall structure. In: FRST 210—Forest biology II. The University of British Columbia, Vancouver. http://frst210.forestry.ubc.ca/lecture/

Alfani F, Gallifuoco A, Saporosi A, Spera A, Cantarella M (2000) Comparison of SHF and SSF processes for the bioconversion of steam-exploded wheat straw. J Ind Microbiol Biotechnol 25(4):184–192

Atanu B, Saha BC, Lawton JW, Shogren RL, Willett JL (2006) Process for obtaining cellulose acetate from agricultural by-products. Carbohydr Polym 64:134–137

Ashokkumar V, Venkatkarthick R, Jayashree S, Chuetor S, Dharmaraj S, Kumar G et al (2022) Recent advances in lignocellulosic biomass for biofuels and value-added bioproducts—a critical review. Bioresour Technol 344:126195

Azuma J, Tanaka F, Koshijima T (1984) Enhancement of enzymic susceptibility of lignocellulosic wastes by microwave irradiation. Hakko Kogaku Zasshi (Japan) 62(4):377–384

Bajpai P, Bajpai P (2016) Structure of lignocellulosic biomass. In: Pretreatment of lignocellulosic biomass for biofuel production. Springer, Singapore, pp 7–12

Bak JS (2014) Electron beam irradiation enhances the digestibility and fermentation yield of water-soaked lignocellulosic biomass. Biotechnol Rep 4:30–33

Belkacemi K, Turcotte G, De Halleux D, Savoie P (1998) Ethanol production from AFEX-treated forages and agricultural residues. In: Biotechnology for fuels and chemicals: proceedings of the nineteenth symposium on biotechnology for fuels and chemicals held May 4-8. 1997, at Colorado Springs, Colorado. Humana Press, pp 441–462

Ballerini D, Desmarquest JP, Pourquie J, Nativel F, Rebeller M (1994) Ethanol production from lignocellulosics: large scale experimentation and economics. Bioresour Technol 50(1):17–23

Bhardwaj N, Kumar B, Agrawal K, Verma P (2020) Bioconversion of rice straw by synergistic effect of in-house produced ligno-hemicellulolytic enzymes for enhanced bioethanol production. Bioresour Technol Rep 10:100352

Bhatia SK, Jagtap SS, Bedekar AA, Bhatia RK, Patel AK, Pant D et al (2020) Recent developments in pretreatment technologies on lignocellulosic biomass: effect of key parameters, technological improvements, and challenges. Bioresour Technol 300:122724

Binod P, Sindhu R, Singhania RR, Vikram S, Devi L, Nagalakshmi S et al (2010) Bioethanol production from rice straw: an overview. Bioresour Technol 101(13):4767–4774

Bouchard J, Nguyen TS, Chornet E, Overend RP (1991) Analytical methodology for biomass pretreatment. Part 2: characterization of the filtrates and cumulative product distribution as a function of treatment severity. Bioresour Technol 36(2):121–131

Branco JEH, Branco DH, de Aguiar EM, Caixeta Filho JV, Rodrigues L (2019) Study of optimal locations for new sugarcane mills in Brazil: application of a MINLP network equilibrium model. Biomass Bioenergy 127:105249

Brownell HH, Yu EKC, Saddler JN (1986) Steam-explosion pretreatment of wood: effect of chip size, acid, moisture content and pressure drop. Biotechnol Bioeng 28(6):792–801

Cavka A, Jönsson LJ (2013) Detoxification of lignocellulosic hydrolysates using sodium borohydride. Bioresour Technol 136:368–376

Chandra RP, Bura R, Mabee WE, Berlin DA, Pan X, Saddler JN (2007) Substrate pretreatment: the key to effective enzymatic hydrolysis of lignocellulosics? In: Olsson L (ed) Biofuels, Advances in biochemical engineering/biotechnology, vol 108. Springer, Berlin, pp 67–93

Chang VS, Holtzapple MT (2000) Fundamental factors affecting biomass enzymatic reactivity. Appl Biochem Biotechnol 84–86:5–37

Claassen PAM, Van Lier JB, Lopez Contreras AM, Van Niel EWJ, Sijtsma L, Stams AJM et al (1999) Utilisation of biomass for the supply of energy carriers. Appl Microbiol Biotechnol 52: 741–755

Cywar RM, Rorrer NA, Hoyt CB, Beckham GT, Chen EYX (2022) Bio-based polymers with performance-advantaged properties. Nat Rev Mater 7(2):83–103

Drummond ARF, Drummond IW (1996) Pyrolysis of sugar cane bagasse in a wire-mesh reactor. Ind Eng Chem Res 35(4):1263–1268

Dubey P, Yousuf O (2021) An overview of fruit by-products valorization: a step towards sustainable utilization. Ind J Pure App Biosci 9(1):46–55

Farooqui A, Tripathi G, Moheet K, Dubey P, Ahmad S, Husain A, Shamim A, Mahfooz S (2021) Algal biomass: potential renewable feedstock for bioenergy production. In: Srivastava M, Srivastava N, Singh R (eds) Bioenergy research: integrative solution for existing roadblock. Springer, Singapore, pp 85–113

Gadde B, Bonnet S, Menke C, Garivait S (2009a) Air pollutant emissions from rice straw open field burning in India, Thailand and The Philippines. Environ Pollut 157(5):1554–1558
Gadde B, Menke C, Wassmann R (2009b) Rice straw as a renewable energy source in India, Thailand, and The Philippines: overall potential and limitations for energy contribution and greenhouse gas mitigation. Biomass Bioenergy 33(11):1532–1546
Galbe M, Zacchi G (2007) Pretreatment of lignocellulosic materials for efficient bioethanol production. Adv Biochem Eng Biotechnol 108:41–65
Gibson LJ (2012) The hierarchical structure and mechanics of plant materials. J R Soc Interface 9(76):2749–2766
Gollapalli LE, Dale BE, Rivers DM (2002) Predicting digestibility of ammonia fiber explosion (AFEX)-treated rice straw. Appl Biochem Biotechnol 98:23–35
Griebl, A., Lange, T., Weber, H., Milacher, W., & Sixta, H. (2005). Xylo-oligosaccharide (XOS) formation through hydrothermolysis of xylan derived from viscose process. In Macromolecular symposia (232, 1, 107-120). Weinheim: Wiley
Gu F, Wang W, Jing L, Jin Y (2013) Sulfite–formaldehyde pretreatment on rice straw for the improvement of enzymatic saccharification. Bioresour Technol 142:218–224
Gupte AP, Basaglia M, Casella S, Favaro L (2022) Rice waste streams as a promising source of biofuels: feedstocks, biotechnologies and future perspectives. Renew Sust Energ Rev 167: 112673
Hadar Y (2013) Sources for lignocellulosic raw materials for the production of ethanol. In: Faraco V (ed) Lignocellulose conversion: enzymatic and microbial tools for bioethanol production. Springer, Berlin, pp 21–38
Harris D, DeBolt S (2010) Synthesis, regulation and utilization of lignocellulosic biomass. Plant Biotechnol J 8(3):244–262
Hessien MM, Rashad MM, Zaky RR, Abdel-Aal EA, El-Barawy KA (2009) Controlling the synthesis conditions for silica nanosphere from semi-burned rice straw. Mater Sci Eng B 162(1):14–21
Hideno A, Inoue H, Tsukahara K, Fujimoto S, Minowa T, Inoue S et al (2009) Wet disk milling pretreatment without sulfuric acid for enzymatic hydrolysis of rice straw. Bioresour Technol 100(10):2706–2711
Hsu TA (2018) Pretreatment of biomass. In: Wyman CE (ed) Handbook on bioethanol, production and utilization: pretreatment of biomass. Taylor and Francis, Washington DC, pp 179–212
Iyer PV, Wu ZW, Kim SB, Lee YY (1996) Ammonia recycled percolation process for pretreatment of herbaceous biomass. Appl Biochem Biotechnol 57:121–132
Jiménez L, Sánchez IY, López F (1990) Characterization of Spanish agricultural residues with a view to obtaining cellulose pulp. TAPPI J 73(8):173–176
Jin S, Chen H (2006) Superfine grinding of steam-exploded rice straw and its enzymatic hydrolysis. Biochem Eng J 30(3):225–230
Jönsson LJ, Alriksson B, Nilvebrant NO (2013) Bioconversion of lignocellulose: inhibitors and detoxification. Biotechnol Biofuels 6(1):1–10
Jørgensen H (2009) Effect of nutrients on fermentation of pretreated wheat straw at very high dry matter content by Saccharomyces cerevisiae. Appl Biochem Biotechnol 153:44–57
Kadam KL, Forrest LH, Jacobson WA (2000) Rice straw as a lignocellulosic resource: collection, processing, transportation, and environmental aspects. Biomass Bioenergy 18(5):369–389
Kádár Z, Szengyel Z, Réczey K (2004) Simultaneous saccharification and fermentation (SSF) of industrial wastes for the production of ethanol. Ind Crop Prod 20(1):103–110
Khan JS (2012) Analysis of Indian agricultural web portals for disseminating information to farmers. Allana Manage J Res Pune 2:67–70
Kim S, Dale BE (2004) Global potential bioethanol production from wasted crops and crop residues. Biomass Bioenergy 26(4):361–375
Kim S, Holtzapple MT (2005) Lime pretreatment and enzymatic hydrolysis of corn stover. Bioresour Technol 96(18):1994–2006

Knappert D, Grethlein H, Converse A (1980) Partial acid hydrolysis of cellulosic materials as a pretreatment for enzymatic hydrolysis. Biotechnol Bioeng 22(7):1449–1463

Koul B, Yakoob M, Shah MP (2022) Agricultural waste management strategies for environmental sustainability. Environ Res 206:112285

Kumakura M, Kaetsu I (1983) Effect of radiation pretreatment of bagasse on enzymatic and acid hydrolysis. Biomass 3(3):199–208

Kumar D, Murthy GS (2011) Impact of pretreatment and downstream processing technologies on economics and energy in cellulosic ethanol production. Biotechnol Biofuels 4:1–19

Kumar B, Bhardwaj N, Agrawal K, Chaturvedi V, Verma P (2020) Current perspective on pretreatment technologies using lignocellulosic biomass: an emerging biorefinery concept. Fuel Process Technol 199:106244

Lee J (1997) Biological conversion of lignocellulosic biomass to ethanol. J Biotechnol 56(1):1–24

Lewandowski I, Scurlock JM, Lindvall E, Christou M (2003) The development and current status of perennial rhizomatous grasses as energy crops in the US and Europe. Biomass Bioenergy 25(4): 335–361

Limayem A, Ricke SC (2012) Lignocellulosic biomass for bioethanol production: current perspectives, potential issues and future prospects. Prog Energy Combust Sci 38(4):449–467

Mat Aron NS, Khoo KS, Chew KW, Show PL, Chen WH, Nguyen THP (2020) Sustainability of the four generations of biofuels—a review. Int J Energy Res 44(12):9266–9282

Menon V, Rao M (2012) Trends in bioconversion of lignocellulose: biofuels, platform chemicals & biorefinery concept. Prog Energy Combust Sci 38(4):522–550

Michelin M, Polizeli MLTM, Ruzene DS, Silva DP, Teixeira JA (2013) Application of lignocelulosic residues in the production of cellulase and hemicellulases from fungi. In: Polizeli MLTM, Rai M (eds) Fungal enzymes. CRC Press, Boca Raton, pp 31–64

Mielenz JR (2001) Ethanol production from biomass: technology and commercialization status. Curr Opin Microbiol 4(3):324–329

Mood SH, Golfeshan AH, Tabatabaei M, Jouzani GS, Najafi GH, Gholami M, Ardjmand M (2013) Lignocellulosic biomass to bioethanol, a comprehensive review with a focus on pretreatment. Renew Sust Energ Rev 27:77–93

Mosier N, Wyman C, Dale B, Elander R, Lee YY, Holtzapple M, Ladisch M (2005) Features of promising technologies for pretreatment of lignocellulosic biomass. Bioresour Technol 96(6): 673–686

Mussatto SI, Dragone G, Guimarães PM, Silva JPA, Carneiro LM, Roberto IC et al (2010) Technological trends, global market, and challenges of bio-ethanol production. Biotechnol Adv 28(6):817–830

Nabarlatz D, Ebringerová A, Montané D (2007) Autohydrolysis of agricultural by-products for the production of xylo-oligosaccharides. Carbohydr Polym 69(1):20–28

Pan X, Arato C, Gilkes N, Gregg D, Mabee W, Pye K et al (2005) Biorefining of softwoods using ethanol organosolv pulping: preliminary evaluation of process streams for manufacture of fuel-grade ethanol and co-products. Biotechnol Bioeng 90(4):473–481

Pandey A (2008) Handbook of plant-based biofuels. CRC Press, Boca Raton

Parajó JC, Garrote G, Cruz JM, Dominguez H (2004) Production of xylooligosaccharides by autohydrolysis of lignocellulosic materials. Trends Food Sci Technol 15(3–4):115–120

Park JY, Kanda E, Fukushima A, Motobayashi K, Nagata K, Kondo M et al (2011) Contents of various sources of glucose and fructose in rice straw, a potential feedstock for ethanol production in Japan. Biomass Bioenergy 35(8):3733–3735

Perlack RD, Wright LL, Turhollow AF, Graham RL, Stokes BJ, Erbach DC (2005) Biomass as feedstock for a bioenergy and bioproducts industry: the technical feasibility of a billion-ton annual supply. DTIC document. U.S. Department of Energy and U.S. Department of Agriculture, Washington, DC

Prasad A, Ross A, Rosenberg P, Dye C (2016) A world of cities and the end of TB. Trans R Soc Trop Med Hyg 110(3):151–152

Rahmati S, Doherty W, Dubal D, Atanda L, Moghaddam L, Sonar P et al (2020) Pretreatment and fermentation of lignocellulosic biomass: reaction mechanisms and process engineering. React Chem Eng 5(11):2017–2047

Ravindran R, Jaiswal AK (2016) A comprehensive review on pre-treatment strategy for lignocellulosic food industry waste: challenges and opportunities. Bioresour Technol 199:92–102

Rezania S, Oryani B, Cho J, Talaiekhozani A, Sabbagh F, Hashemi B et al (2020) Different pretreatment technologies of lignocellulosic biomass for bioethanol production: an overview. Energy 199:117457

Robak K, Balcerek M (2020) Current state-of-the-art in ethanol production from lignocellulosic feedstocks. Microbiol Res 240:126534

Robinson P (2006) University of California Davis, personal communication

Saha BC (2003) Hemicellulose bioconversion. J Ind Microbiol Biotechnol 30(5):279–291

Saha BC, Iten LB, Cotta MA, Wu YV (2005) Dilute acid pretreatment, enzymatic saccharification and fermentation of wheat straw to ethanol. Process Biochem 40(12):3693–3700

Saini JK, Saini R, Tewari L (2015) Lignocellulosic agriculture wastes as biomass feedstocks for second-generation bioethanol production: concepts and recent developments. 3 Biotech 5:337–353

Sánchez C (2009) Lignocellulosic residues: biodegradation and bioconversion by fungi. Biotechnol Adv 27(2):185–194

Sarkar N, Ghosh SK, Bannerjee S, Aikat K (2012) Bioethanol production from agricultural wastes: an overview. Renew Energy 37(1):19–27

Saxena RC, Adhikari DK, Goyal HB (2009) Biomass-based energy fuel through biochemical routes: a review. Renew Sust Energ Rev 13(1):167–178

Schmidt AS, Thomsen AB (1998) Optimization of wet oxidation pretreatment of wheat straw. Bioresour Technol 64(2):139–151

Schmitt E, Bura R, Gustafson R, Cooper J, Vajzovic A (2012) Converting lignocellulosic solid waste into ethanol for the state of Washington: an investigation of treatment technologies and environmental impacts. Bioresour Technol 104:400–409

Schutyser W, Renders AT, Van den Bosch S, Koelewijn SF, Beckham GT, Sels BF (2018) Chemicals from lignin: an interplay of lignocellulose fractionation, depolymerisation, and upgrading. Chem Soc Rev 47(3):852–908

Singh A, Basak P (2019) Economic and environmental evaluation of rice straw processing technologies for energy generation: a case study of Punjab, India. J Clean Prod 212:343–352

Singh A, Pant D, Korres NE, Nizami AS, Prasad S, Murphy JD (2010) Key issues in life cycle assessment of ethanol production from lignocellulosic biomass: challenges and perspectives. Bioresour Technol 101(13):5003–5012

Singh R, Tiwari S, Srivastava M, Shukla A (2013) Performance study of combined microwave and acid pretreatment method for enhancing enzymatic digestibility of rice straw for bioethanol production. Plant Knowledge J 2(4):157–162

Singh R, Shukla A, Tiwari S, Srivastava M (2014a) A review on delignification of lignocellulosic biomass for enhancement of ethanol production potential. Renew Sust Energ Rev 32:713–728

Singh R, Tiwari S, Srivastava M, Shukla A (2014b) Microwave assisted alkali pretreatment of rice straw for enhancing enzymatic digestibility. J Energy 2014:1–7

Singh R, Srivastava M, Shukla A (2016) Environmental sustainability of bioethanol production from rice straw in India: a review. Renew Sust Energ Rev 54:202–216

Singh P, Dubey P, Younis K, Yousuf O (2022) A review on the valorization of coconut shell waste. Biomass Conv Bioref. https://doi.org/10.1007/s13399-022-03001-2

Sticklen M (2006) Plant genetic engineering to improve biomass characteristics for biofuels. Curr Opin Biotechnol 17(3):315–319

Sulbarán-de-Ferrer B, Aristiguieta M, Dale BE, Ferrer A, Ojeda-de-Rodriguez G (2003) Enzymatic hydrolysis of ammonia-treated rice straw. Appl Biochem Biotechnol 105:155–164

Sun Y, Cheng J (2002) Hydrolysis of lignocellulosic materials for ethanol production: a review. Bioresour Technol 83(1):1–11

Taherzadeh MJ, Niklasson C (2004) Ethanol from lignocellulosic materials: pretreatment, acid and enzymatic hydrolyses, and fermentation. In: Saha BC, Hayashi K (eds) Lignocellulose biodegradation. ACS Publications, Washington, DC, pp 49–68

Takacs E, Wojnarovits L, Földváry C, Hargittai P, Borsa J, Sajo I (2000) Effect of combined gamma-irradiation and alkali treatment on cotton–cellulose. Radiat Phys Chem 57(3–6): 399–403

Teymouri F, Laureano-Perez L, Alizadeh H, Dale BE (2005) Optimization of the ammonia fiber explosion (AFEX) treatment parameters for enzymatic hydrolysis of corn stover. Bioresour Technol 96(18):2014–2018

Tran TTA, Le TKP, Mai TP, Nguyen DQ (2019) Bioethanol production from lignocellulosic biomass. In: Yun Y (ed) Alcohol fuels—current technologies and future prospect. IntechOpen, London, pp 1–14

Tripathi G, Jamal A, Jamal T, Faiyaz M, Farooqui A (2022) Phyco-nanotechnology: an emerging nanomaterial synthesis method and its applicability in biofuel production. In: Srivastava M, Malik MA, Mishra P (eds) Green nano solution for bioenergy production enhancement. Springer Nature, Singapore, pp 169–200

Tye YY, Lee KT, Abdullah WNW, Leh CP (2016) The world availability of non-wood lignocellulosic biomass for the production of cellulosic ethanol and potential pretreatments for the enhancement of enzymatic saccharification. Renew Sust Energ Rev 60:155–172

Van Dyk JS, Pletschke B (2012) A review of lignocellulose bioconversion using enzymatic hydrolysis and synergistic cooperation between enzymes—factors affecting enzymes, conversion and synergy. Biotechnol Adv 30(6):1458–1480

Vazquez M, Oliva M, Tellez-Luis SJ, Ramírez JA (2007) Hydrolysis of sorghum straw using phosphoric acid: evaluation of furfural production. Bioresour Technol 98(16):3053–3060

Volynets B, Ein-Mozaffari F, Dahman Y (2017) Biomass processing into ethanol: pretreatment, enzymatic hydrolysis, fermentation, rheology, and mixing. Green Process Synth 6(1):1–22

Watanabe A, Katoh K, Kimura M (1993) Effect of rice straw application on CH_4 emission from paddy fields: II. Contribution of organic constituents in rice straw. Soil Sci Plant Nutr 39(4): 707–712

Wyman C (1996) Handbook on bioethanol: production and utilization. CRC Press, Boca Raton

Xiong J, Ye J, Liang WZ, Fan PM (2000) Influence of microwave on the ultrastructure of cellulose I. J South China Univ Technol 28(1):84–89

Yao RS, Hu HJ, Deng SS, Wang H, Zhu HX (2011) Structure and saccharification of rice straw pretreated with sulfur trioxide micro-thermal explosion collaborative dilutes alkali. Bioresour Technol 102(10):6340–6343

Zabed H, Faruq G, Sahu JN, Azirun MS, Hashim R, Nasrulhaq Boyce A (2014) Bioethanol production from fermentable sugar juice. Sci World J 2014:957102

Zabed H, Sahu JN, Boyce AN, Faruq G (2016) Fuel ethanol production from lignocellulosic biomass: an overview on feedstocks and technological approaches. Renew Sust Energ Rev 66:751–774

Zhao X, Zhang L, Liu D (2012) Biomass recalcitrance. Part I: the chemical compositions and physical structures affecting the enzymatic hydrolysis of lignocellulose. Biofuels Bioprod Biorefin 6(4):465–482

Chapter 6
Nanomaterial Fabrication from Paddy Straw for Bioethanol Production

Manmeet Kaur, Diksha Singla, Harinderjeet Kaur, Sandeep Kaur, Tarun Sharma, and Ravish Choudhary

Abstract The depletion of natural resources and the relatively small carbon footprint they generate are the main reasons for the push for renewable energy sources. Because they are renewable and emit little carbon dioxide, biofuels are considered the best prospects for use as alternative energy sources. Furthermore, the lack of effective processes and technologies makes large-scale bioethanol synthesis time-consuming. Pretreatment of the rigid and inflexible agro-industrial waste to release sugars is a mandatory step for ethanol production. Previously, this was achieved through chemical, biological, and physical processes that required expensive solvents and chemicals, large amounts of energy, effective organisms, and high temperatures and pressures to release the tightly bound configuration of the substrates. The production of unfavorable and inhibitory compounds as well as equipment degradation is caused by these expensive and complex technologies. Nanotechnology has been extensively studied for its potential to improve biofuel production methods due to its superior adsorption, catalysis, and reactivity capabilities. The role of these advanced nanotechnologies and associated nanoscale elements in biomass conversion as well as their impact on yield and bioethanol production processes are elaborated in this chapter.

M. Kaur (✉)
Department of Microbiology, Punjab Agricultural University, Ludhiana, Punjab, India
e-mail: manmeet1-mb@pau.edu

D. Singla
Department of Biochemistry, Punjab Agricultural University, Ludhiana, Punjab, India

H. Kaur
Department of Food Science, Rayat Bahra University, Mohali, Punjab, India

S. Kaur
Department of Entomology, Punjab Agricultural University, Ludhiana, Punjab, India

T. Sharma
Department of Agronomy, CSK Himachal Pradesh Agricultural University, Palampur, Himachal Pradesh, India

R. Choudhary
Division of Seed Science and Technology, Indian Agricultural Research Institute, Delhi, India

N. Srivastava et al. (eds.), *Paddy Straw Waste for Biorefinery Applications*, Clean Energy Production Technologies, https://doi.org/10.1007/978-981-99-8224-0_6

Keywords Paddy straw · Bioethanol · Nanoparticles · Nanocomposites · Nanocatalyst · Renewable energy

6.1 Introduction

Over-reliance on conventional energy sources inevitably leads to rapid depletion of renewable energy supplies. In addition, it causes major environmental problems such as climate change and loss of ecological integrity. Because of these dire circumstances, scientists are exploring the prospects of viable energy substitutes. Solar energy, wind power, biomass, and other forms of renewable energy are often viewed as viable alternatives to traditional energy sources. However, because biomass energy production is so widespread worldwide, it is guaranteed to take the top spot among these forms of energy (Antunes et al. 2017).

Bioethanol is viewed as a safe and environmentally friendly substitute for gasoline as an independent fuel or as a fuel alongside petrol or diesel. The application of this substance is anticipated to entirely halt the emission of SO_2, which is responsible for acidic rain, and reduce the greenhouse gas emissions from cellulosic ethanol and maize ethanol by 80% and around 20%, respectively. Enzymatic hydrolysis of sugars, pretreatment of lignin-rich biomass, and fermentation of hydrolysate to produce ethanol and other significant biochemicals are the three steps in the production of bioethanol. To conduct simultaneous saccharification (SSF), simultaneous saccharification and co-fermentation (SSCF), and consolidated bioprocessing (CBP), stages (ii) and (iii) may occasionally be combined. The essential steps of the process of converting rice straw into bioethanol are schematically depicted in Fig. 6.1.

The lignin released by the pretreatment can also be used to produce bioenergy and other products with significant value.

A range of chemical, physical, physicochemical, and biological pretreatment techniques are routinely used to process agro-industrial residues. Most of these methods are considered fascinating, but their common application is limited by certain drawbacks. Therefore, nanotechnology can provide novel solutions to the research need to develop innovative alternatives to current pretreatment methods as it can alter the structure of raw materials and biocatalysts used in the generation of bioethanol. Nanotechnology can intervene in this situation and remove the significant obstacles that are ruling in the industries. In recent years, nanotechnology has grown rapidly worldwide, significantly advancing and exploiting an increasing number of products in various sectors (Leo and Singh 2018). In light of this, this chapter examines cutting-edge nanotechnology techniques that are enormously utilized in the production of bioethanol.

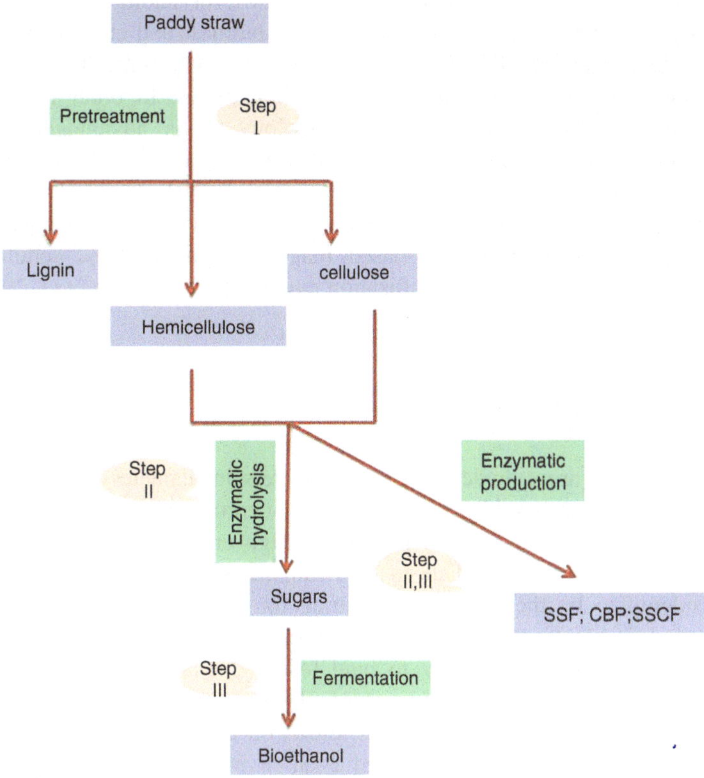

Fig. 6.1 Schematic illustration of the key phases in the synthesis of bioethanol from paddy straw

6.2 Novel Nanomaterials Utilized in Bioethanol Production

6.2.1 Magnetic Nanoparticles (MNPs)

Magnetic domain and paramagnetic atom are two analogies that produce nanoparticles through the passage of magnetic field. The magnetic properties of nanoparticles arise from both their intrinsic properties and the way they interact with each other (Singh et al. 2020). Due to their tiny dimensions, excellent quantum properties, and effective surface area, magnetic nanoparticles are capable of transporting other molecules. Magnetic nanoparticles' primary drawback is that they are inherently unstable over long periods of time since they assemble and emit energy due to their substantial high surface (Lai et al. 2018). Due to their robust chemical activity, the magnetic nanoparticles are also easily oxidized and dispersed, compromising their magnetic properties. To protect the magnetic nanoparticles from chemical degradation during and after the production phase, techniques such as coating or modification of layers can be recommended (Zhong and Sels 2018). According to Kim and Lee (2016), the addition of cobalt ferrite silicate nanoparticles increased the production of biomass and ethanol by 227.01% and 214.1%, respectively. Biocatalyst reuse and recovery becomes less complicated by immobilizing biocatalysts on magnetic nanoparticles. In order to produce a practical and ecologically acceptable technology, it is advantageous to use nanobiocatalysts for the catalysis of agro-industrial biomass (Rai et al. 2019). The prospect of reducing the amount of enzyme required for cellulose hydrolysis comes from the improved stability of β-glucosidase immobilized on magnetic nanoparticles (Park et al. 2018). Using lignin-based magnetic nanoparticles, cellulase was immobilized and resulted in immobilization rate of 55.52% (Dong et al. 2019).

6.2.2 Carbon Nanotubes (CNTs)

Carbon nanotubes are cylinders a few nanometers in diameter made of graphite layers that exhibit remarkable mechanical and biocompatible properties. The large surface area of carbon nanotubes guarantees high enzyme loading capacity and minimal diffusion resistance. Due to their numerous beneficial properties, CNTs are considered to be the best nano-substrate for use in enzyme immobilization (Bianco et al. 2011). Carbon nanotubes are regarded as an essential facet in biofuel applications due to their capacity to bind to enzyme-active sites for immediate transfer of electrons and the presence of electroactive zone that promotes the quantity of enzymes and other redox chemicals on their surface (Rai et al. 2016). Multiwalled carbon nanotubes were used as an environmentally friendly and long-lasting catalyst to rapidly convert ethanol into acetaldehyde. With an ethanol conversion of about 60% and an acetaldehyde selectivity of 93%, the improved carbon nanotube catalyst enabled a remarkable level of reactivity (Wang et al. 2016a). A pervaporation membrane bioreactor produced the ethanol with highest productivity (0.51 g/h) (Amrei et al. 2020). Fathy (2017) claims that hydrothermally treated rice straw with nickel nitrate salt in the presence of ferrocene was used to produce carbon nanotubes. The carbon nanotube and zinc oxides/porous silica layer exhibited the greatest sensitivity to ethanol gas, with a peak value of 2.004. The combination of carbon nanotubes with zinc oxide leads to the formation of novel features and characteristic effects due to the enhanced sensing capabilities of CNTs in identifying various gases at suitable temperature (Hussein et al. 2021).

6.2.3 Mesoporous Silica

Trisunaryanti et al. (2020) claim that the advantages of mesoporous silica for enzyme immobilization include its compatibility, high pore volume, hydrothermal stability, and ease of surface modification. The covalent immobilization of active centers is made possible by the silicon dioxide-based substances, which have a higher proportion of silanol molecules. Such properties are widely embraced to improve catalytic performance in acid-catalyzed chemical reactions in terms of activity, efficiency, and robustness (Bhanja and Bhaumik 2016). According to Lee et al. (2014), 51% fructose production was achieved when cellulose waste was converted utilizing iron oxide-encapsulated mesoporous silica nanoparticles. Chae et al. (2014) discovered that to maximize the catalytic effectiveness of tantalum oxide substrates for the production of butadiene from ethanol and acetaldehyde, the mesopore structure of mesoporous silica samples was less important than the crystal and pore size. The ethanol space-time yield of the cobalt-copper catalyst doped on mesoporous silica was reported to be 0.23 mmol/(gmetal/h) (Wang et al. 2022a, 2022b)

6.2.4 Zeolites

In petrochemical and fine industrial plants, zeolites are often used as a primary heterogeneous catalyst due to their crystalline inorganic framework, which gives them exceptional heat stability. In addition, zeolites have well-organized micropores and strong lewis and bronsted acid centers. It is important to emphasize that zeolites have limited applications in liquid due to disadvantages such as their sensitivity toward hot water. Zeolite active site functions as carbonium ion and can be used for the combustion of biomass (Markandan and Chai 2022). The porosity of catalysts constructed on the ZSM-5 platform can be altered by varying the concentration of silicon or aluminum in the zeolite. According to Jorge et al. (2017), the lead-exchanged catalyst showed more efficient performance in converting furfural to isopropyl levulinate.

6.2.5 Metal Organic Frameworks (MOFs)

MOFs are exceptionally effective due to their exceptional high surface area. Scientists discovered that several metal-organic frameworks functionalized with sulfonic acids were used as a catalyst and playing an crucial role in alteration of fructose into 5-hydroxymethylfurfural. Fructose conversion with highest yield was reported to be 90%. Since it is considered a heterogeneous catalyst and is easy to utilize and reuse, a solid acid catalyst composed of a metal-organic framework is promising for exploiting the carbohydrate content of biomass (Chen et al. 2014). According to Li et al. (2019), MOFs quantitatively determine the water content of ethanol (dry) and bioethanol with a yield ranged between 90 and 107%. The addition of MOF cocatalysis resulted in the most significant amounts of reducing sugar (390 mg/g) and ethanol production (70%) (Ahmed et al. 2019). ZIF-71 (chlorobromide)-SE acts as a favorable material for biofuel purification applications due to its enhanced hydrophobic nature and maximum adsorption ability (Tiempos-Flores et al. 2022).

According to de Lima et al. (2013), the metal-organic framework Zn_2 $(BDC)_2(TED)$ is extremely specific for the adhesion of alcohols and water and also functions as filtration membrane during the ethanol production. Adsorbent-mediated ethanol adsorption from water and methanol is a successful and promising method for producing ethanol suitable for use in fuels (Li et al. 2022).

6.3 Advanced Approaches: Nanotech Biofuel World

The latest methods for pretreatment of lignocellulosic biomass before conversion to simpler sugars and ethanol are discussed in the following sections

6.3.1 Nano-Shear Hybrid Alkaline Technique (NSHA)

During shearing at increased speed and at comparatively low temperatures, chemical reagents for this process are coupled in special reactors, so-called nanomixers. Wang et al. (2013) claim that it is mainly used to remove lignin units from lignocellulosic biomass after short-term treatment. These studies showed that NSHA pretreatment effectively removed lignin and hemicelluloses, which contributed to the degradation of the cellulose nanostructure. This technology reported that complex parts of the lignocellulosic biomass (lignin and hemicellulose) were eliminated, but the cellulose remained at par around 80% (Arora et al. 2020). The synergistic action of cellulase enzymes reduced biomass resistance and generated nanoscale polysaccharide aggregation and easily converted into simple sugars by this advanced technology. This mechanism offers advantages for introducing high shear forces into various chemical products and pretreatment processes due to the synergistic effects of high shear rates (de Oliveira et al. 2017).

6.3.2 Nanomaterials

Nanomaterials can have either immediate or subsequent impacts on bioethanol production methods (Verma et al. 2013). The main applications of these nanoparticles are the effective immobilization of these enzymes and the enzymatic hydrolysis of lignin-based biomass. Laccases, hemicellulases, and other important lignocellulosic degradants are immobilized in metal oxide nanoparticle matrices (Rai et al. 2016). These nanomaterial-assisted enzymes, also called nanocatalysts, have gained popularity due to their improved effectiveness. Pretreatment, catalytic hydrolysis, saccharification, and purification are the four main categories into which nanomaterials are widely integrated. An efficient nano-biocatalyst was demonstrated using methyl-functionalized cobalt ferrite silica, as it produces the most ethanol

(127.1%) among all materials evaluated (Bosu et al. 2022). According to Abdelkareem et al. (2020), cobalt was deposited on the surface of nickel foam and showed significant activity as an electrode for ethanol oxidation. The nanomaterial CZ75 can be used as a renewable functional catalyst for the conversion of ethanol (Mahmoud 2014).

6.3.3 Nanomagnetic Nanocatalyst

According to Sirajunnisa and Surendhiran (2016), magnetic nanoparticles have a number of industrial applications due to their magnetic properties for adhering to substrates, their lower toxicity concerns, their nanoscale size, and their potential to accelerate chemical reactions. Kim and Lee (2016) studied the effects of methyl-functionalized silica and cobalt ferrite-silica nanoparticles on increasing bioethanol production after syngas-based fermentation of *Clostridium ljungdahlii*. OPO_3H groups were immobilized on an iron oxide nanocellulose surface to prepare a bio-based magnetic nanocatalyst (Salehi and Mirjalili 2017). This study demonstrates the high yield and recyclable catalytic properties of this enzyme. Cellulase was immobilized on magnetic nanoparticles using ultrasonic hyperactivation, which increased the catalytic activity of the enzyme by 3.6-fold (Ladole et al. 2017). Copper oxide nanoarchitectures were developed using nanoporous copper substrates immersed in anhydrous ethanol (Dan et al. 2018).

6.3.4 Nanofiltration

One of the key factors affecting the production and quality of bioethanol is the presence of chemical and microbiological pollutants. Due to their low energy requirements, easy operation, cheap labor costs, and small working space, membrane separation technologies such as nanofiltration have attracted attention in the bioethanol purification industry. Nanofiltration could also be used during fermentation to concentrate the sugars in the mixture and remove possible yeast-based fermentation inhibitors. Nanofiltration is one such high-pressure membrane filtration method that is frequently used to separate bioethanol (Kang et al. 2014). A combination of forward osmosis (FO) and nanofiltration (NF) techniques have been developed to enhance the bioethanol production by utilizing xylose-assimilating *S. cerevisiae* (Shibuya et al. 2017). Additionally, this hybrid system has been proven to efficiently eliminate known fermentation inhibitors. The effective separation of bioethanol from pretreated biomass has been significantly improved by these hybrid nanofiltration technologies.

6.3.5 Nanofibers

To cross-link nanofibers, β-glucosidase enzymes can be administered in a placed form (Lee et al. 2010). This enzyme will be extremely beneficial, especially in eliminating the inhibition brought on by too much cellobiose. This method allowed the enzyme to keep 90% of its activity with the potential of higher enzyme retention owing to the usage of magnetic nanofibers for reuse and recycling. According to Wang et al. (2015), the Yb-doped tin oxide hollow nanofiber-based sensors had the greatest effectiveness for the detection of ethanol, as well as exceptional selectivity, high responsiveness, and fast response times. The ethanol sensor made of zinc oxide-tin oxide nanofibers has a distinctive oxygen vacancy, hollow structure, and n-n heterojunction, which enhance sensing performance and reduce recovery time (Li et al. 2015).

6.3.6 Nanocatalyst

As shown in Fig. 6.2, these enzymes can be immobilized on nanoparticles through various techniques. These enzymes immobilized on nanoparticles are called nanocatalysts. Cellulase enzyme has the ability to breakdown the complex lignocellulosic substrate into simple sugars. Therefore, using immobilized nanocellulases might elevate the enzymatic capacity for reuse and recuperation rate (Rai et al. 2016). Numerous biocatalysts, including cellulase, xylanase, cellobiose, β-glucosidase, and laccase, can be immobilized on nanomaterials to improve the efficiency and catalytic rate of enzymes (Rai et al. 2019). Ni-Pd/C is more actively

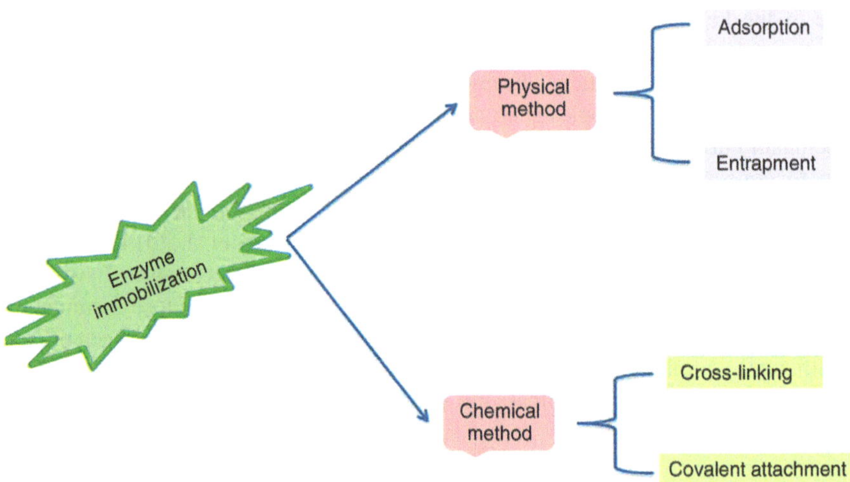

Fig. 6.2 Mechanisms involved in immobilization of enzymes onto nanoparticles

involved in the ethanol oxidation process than Pd/C. The palladium-nickel surface also reorganizes the catalyst surface and improves the arrangement of nickel and palladium sites for effective ethanol oxidation (Obradovic et al. 2016). According to the research of Wu et al. (2017), engineered copper-nickel-magnesium-aluminum oxide nanocatalysts showed maximum level of activity and selectivity for the hydrogenation process leading to the conversion of furfural into tetrahydrofurfuryl during ethanol production. Palladium-silver/multiwalled carbon nanotube electrocatalysts exhibit excellent electro-chemical activity for the ethanol oxidation (Adam et al. 2020). Most active catalysts are required to break down the carbon-linked bonds during ethanol formation (Bai et al. 2019).

6.3.7 Nanosensors

An essential step in the production of bioethanol is the identification of ethanol through nanosensors. An ethanol gas sensor is built from branched $TiO_2/Ag_{0.35}V_2O_5$ nanoheterostructures (Wang et al. 2016b). This sensor is significantly different from traditional pure tin oxide nanofiber biosensors due to its fast response time, excellent selectivity, and high sensitivity. According to Wang et al. (2022a, 2022b) in ethanol, a Ce-doped (2%) device had a response time of 69.2, which is 2.1 times higher than the response time of a tin oxide. At room temperature, a zinc oxide nanoparticle-based surface acoustic wave ethanol gas sensor works with great sensitivity and maintains its reliability over a short period of time (Wu et al. 2012). Tin oxide sensors coated with iron oxide nanoparticles have been reported to be 1.5 to 7.1 times more sensitive to ethanol gas (Choi et al. 2017). In terms of reliability, instantaneous response, sensitivity, and durability, gold nanoparticles were crucial in improving the analytical effectiveness of the ethanol sensor (Madaci et al. 2023). Tin sulfide nanoparticles are more susceptible to sensing ethanol gas due to their chain-like structure and superior stoichiometry (Rana et al. 2019).

6.4 Nano-Based Pretreatment of Paddy Straw

The crucial step in the production of bioethanol is pretreatment, which helps to modify the configuration of rice straw and facilitates the ability of cellulases to bind to carbohydrates and hydrolyze them into simple sugars (Kaur et al. 2023). However, the most effective pretreatment approaches should take a number of things into account, including:

- Be able to hydrolyze sugars
- Reduce sugar loss
- Prevent the production of by-products
- Restrict fermentation processes
- Must be economically viable

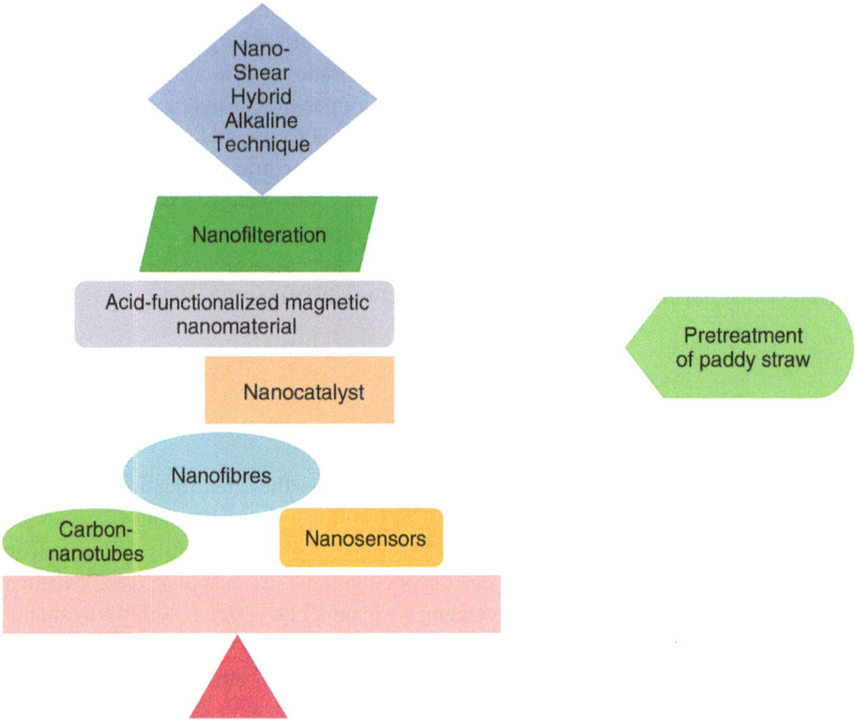

Fig. 6.3 Agro-industrial pretreatment nanotech advanced techniques

The ability to magnetically remove and recover nanoparticles from the medium reduces the cost of the manufacturing process and enables the implementation of pretreatment techniques based on nanotechnology (Fig. 6.3). The structural configuration of the biomass can be changed by simply incorporating the nanoparticles during the pretreatment process. Due to their nanosize, metal nanoparticles may also react more easily with lignocellulosic components and release reducing and total sugars with minimal amount of inhibitor (Razack et al. 2016).

The use of nanobiocatalysts in processes such as acid functionalization and enzyme immobilization is expected to modernize existing techniques as they can be used continuously for numerous hydrolysis reactions (Pena et al. 2012). The use of magnetic nanocatalysts with acid-functionalized enzymes could reduce the complexity of the pretreatment processes since the acid would not need to be used in every cycle. However, since a given enzyme can be used in numerous cycles, the continuous use of nanocatalysts with immobilized enzymes would be more cost-effective. As shown in Fig. 6.4, the release of significant amounts of fermentable sugars could potentially be beneficial. The simpler carbohydrates make up the majority of the pretreatment product. These are either hydrolyzed by acids or converted into sugars and acids by enzymatic activity. Acidic sugar separation is

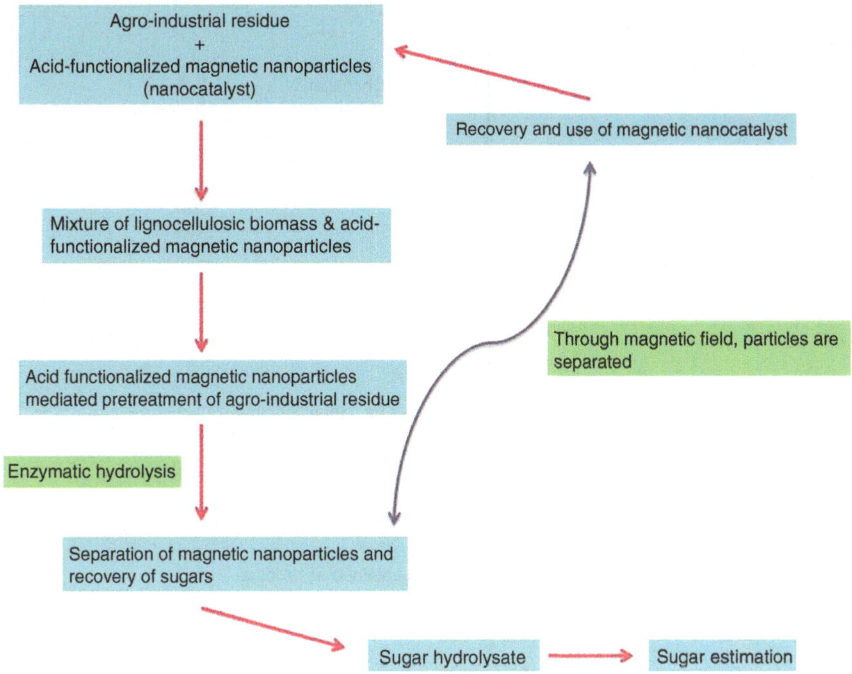

Fig. 6.4 Acid-functionalized magnetic nanoparticles mediated pretreatment of agro-industrial residues

necessary because the sugar component can only be used during fermentation. The most popular and least efficient separation method is chromatographic separation. Pure sugar molecules can only be produced with nanofilters. Nanofilter membrane prevents the acid from passing through while still keeping the sugar molecules in place. The large surface area of the nanomaterials allows them to be linked with the specific amount of enzyme or acid.

6.5 Applicability of Nano-Advanced Techniques in Bioethanol Production

Among liquid biofuels, bioethanol is one of the most important by-products from biorefineries. Due to the numerous advantages that bioethanol offers over traditional fuels, its production has received considerable attention. Various biomass and industrial wastes are the main non-food raw materials for the production of bioethanol (Tropea 2022). Bioethanol production is significantly influenced by nanotechnology (Fig. 6.5). The inclusion of nickel oxide nanoparticles in the pretreatment process increased the yield of bioethanol by 1.6 times and reduced

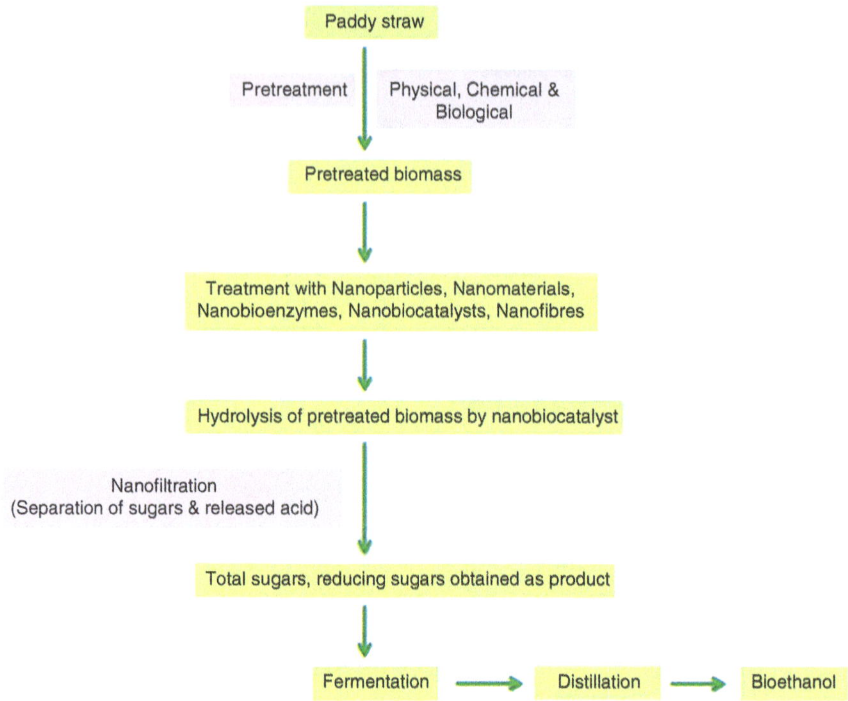

Fig. 6.5 Schematic representation of bioethanol production via nanotechnology

the amount of acetic acid by 2.10 times (Sanusi et al. 2021). ZnO nanoparticles were used to increase the ethanol yield (Gupta and Chundawat 2020). The saccharification yields of alkali-treated rice straw were improved (375.39 mg/gds) by using covalently immobilized magnetic enzyme-nanoparticle complexes (Kumar et al. 2017). Rice straw immobilized after treatment with cellulase was hydrolyzed with a saccharification efficiency of 52.67 percent. A maximum of four saccharification cycles could be performed while still retaining 50.34% of its activity (Kaur et al. 2021). When immobilized on magnesium dioxide nanoparticles, cellulase enzymes can synthesize bioethanol (Cherian et al. 2015).

When immobilized *A. niger* on nickel ferrite and iron oxide and immobilized Cellic Tec2 on nickel ferrite were used to hydrolyze rice straw, the amounts of fermentable sugars produced were 1.89, 1.91, and 7.2 mg/mL, respectively (Nayeri et al. 2023). Cellulase was immobilized using cyclodextrin iron oxide nanoparticles, and rice straw was hydrolyzed using an ionic solvent. The initial response rate increased from 1.63 to 2.74 g/h/L with the immobilized cellulase (Huang et al. 2015). Lin et al. (2016) used a two-step hydrothermal process to fabricate extremely thin zinc oxide nanosheets with uniformly distributed silver nanoparticles to improve surface response for ethanol monitoring. It turned out that silver was used, a precious metal because it is extremely sensitive to gases. The ethanol vapor was released, and the resistance was changed to monitor the activity of the zinc oxide-silver composite.

The resistance was varied using gas sensors that can operate at exceptionally low concentrations of ethanol vapor.

6.6 Significance of Nanotechnology in Riddance of Inhibitors

During the biomass pretreatment phase, numerous inhibitors are produced, including furans, carboxylic acid, and phenolic compounds. To eliminate inhibitory compounds and accelerate the fermentation of hydrolysates, a detoxification step is required. Tin oxide was used in conjunction with nanoparticle catalysts for the photocatalytic degradation of phenolic compounds. However, this method has not been applied to the hydrolyzate by-products of agro-industrial biomass (Rai et al. 2016). Phenol rapidly converts to carboxylic acid when exposed to UV radiation in the presence of iodine-doped tin oxide nanoparticles (Hamdi et al. 2015). When used in biomass pretreatment, nanoparticles significantly improve chemistry at the molecular level and enable special catalyst modification (Razack et al. 2016). Due to their small size, metallic nanoparticles are very effective in disrupting the cell walls of untreated substrates and interact positively with biomolecules to cause the release of sugars (Kushwaha et al. 2018). Phenol is known to photochemically degrade when exposed to UV light and tin dioxide nanoparticles. However, the large band gap of tin oxide makes phenol degradation less efficient under solar radiation (Hamdi et al. 2016).

6.7 Conclusion

Rice straw is one of the major agro-industrial waste which was utilized in the synthesis of environmentally friendly bioethanol. Although there is great interest and potential in exploiting the technology to produce bioethanol from high-yield first- and third-generation feedstocks, the main obstacles are still rising production costs and technological advances. With the help of nanotechnology, these difficulties for the environmentally friendly production of bioethanol could be significantly reduced. The unique properties of nanoparticles, including their high adsorption, rigidity, and reactivity, can lead to significant advances in biomass conversion. The end product and efficiency of biofuels are improved through the immobilization and activation of enzymes, which also increases microbial and catalytic activity. Nanotechnological approaches were used to effectively reduce the stiffness and crystal structure of the polysaccharides during the pretreatment process and ethanol production. There is still much to be done regarding the importance of nanotechnology for the pretreatment of complex and rigid lignocellulosic biomass for enzymatic and biological saccharification. However, further research needs to be conducted on how nanotechnology can be managed, developed, and sustained in biofuel production before its widespread use in biorefineries.

6.8 Future Prospects of Nanotechnology in Bioethanol Production

The incorporation of nanoparticles increases the efficiency of the biofuel production method (Nizami and Rehan 2018). The use of nanoparticles can achieve higher product yields, more effective biomass conversion, and ultimately cheaper costs. There are still some problems to be solved, although nanoparticles can reduce the hurdles in the production of biofuels quite effectively. Finding a greener method to produce more adaptive and selective nanoparticles is the first problem to be solved (Devi et al. 2021). The second research area is the study of nanoparticles to localize reaction intermediates and end products to control, monitor, and optimize the cost of the biofuel production process (Kushwaha et al. 2018). Another difficult problem is the commercialization of nanoparticles. Therefore, strategies for producing effective, affordable, and ecologically harmless nanoparticles must be developed. The next generation of studies should focus on biofuel transportation and storage. Despite the promising future of nanotechnology in biofuel production, more research is still needed to solve the technical problems and make the transition from traditional fossil fuels to environmentally friendly biofuels (Sheikh et al. 2023).

References

Abdelkareem MA, Sayed ET, Alawadhi H, Alami AH (2020) Synthesis and testing of cobalt leaf-like nanomaterials as an active catalyst for ethanol oxidation. Int J Hydrog Energy 45:11–19

Adam AMM, Deng M, Zhu A, Zhang Q, Liu Q (2020) Facile one-step room temperature synthesis of PdAg nanocatalysts supported on multi-walled carbon nanotubes towards electro-oxidation of methanol and ethanol. Electrochim Acta 339:135929

Ahmed IN, Yang XL, Dubale AA, Shao R, Guan RF, Meng X, Xie MH (2019) Zirconium based metal-organic framework in-situ assisted hydrothermal pretreatment and enzymatic hydrolysis of platanus X acerifolia exfoliating bark for bioethanol production. Bioresour Technol 280:13–21

Amrei SS, Asghari M, Esfahanian M, Zahraei Z (2020) Highly selective carbon nanotube-coupled graphene oxide-incorporated polydimethylsiloxane membrane for pervaporative membrane bioreactor ethanol production. J Chem Technol Biotechnol 95:4–13

Antunes FAF, Gaikwad S, Ingle AP, Pandit R, dos Santos JC, Rai M, da Silva SS (2017) Bioenergy and biofuels: nanotechnological solutions for sustainable production. In: Rai M, Da Silva SS (eds) Nanotechnology for bioenergy and biofuel production: green chemistry and sustainable technology. Springer, Cham, pp 3–18

Arora A, Nandal P, Singh J, Verma ML (2020) Nanobiotechnological advancements in lignocellulosic biomass pretreatment. Mater Sci Energy Tech 3:308–318

Bai JD, Liu DD, Yang JF, Chen YP (2019) Nanocatalysts for electrocatalytic oxidation of ethanol. Chem Sustain Energy Mater 12:17–32

Bhanja P, Bhaumik A (2016) Porous nanomaterials as green catalyst for the conversion of biomass to bioenergy. Fuel 185:432–441

Bianco A, Kostarelos K, Prato M (2011) Making carbon nanotubes biocompatible and biodegradable. Chem Commun 47:82–88

Bosu S, Pooja RP, Rajasimman M (2022) Role of nanomaterials in enhanced ethanol production through biological methods—review on operating factors and machine learning applications. Fuel 320:123905

Chae HJ, Kim TW, Moon YK, Kim HK, Jeong KE, Kim CU, Jeong SY (2014) Butadiene production from bioethanol and acetaldehyde over tantalum oxide-supported ordered mesoporous silica catalysts. Appl Catal B Environ 151:596–604

Chen J, Li K, Chen L, Liu R, Huang X, Ye D (2014) Conversion of fructose into 5-hydroxymethylfurfural catalyzed by recyclable sulfonic acid-functionalized metal–organic frameworks. Green Chem 16:2490–2499

Cherian E, Dharmendirakumar M, Baskar G (2015) Immobilization of cellulase onto MnO_2 nanoparticles for bioethanol production by enhanced hydrolysis of agricultural waste. Cuihua Xuebao 36(8):23–29

Choi KS, Park S, Chang SP (2017) Enhanced ethanol sensing properties based on SnO_2 nanowires coated with Fe_2O_3 nanoparticles. Sensors Actuators B Chem 238:71–79

Dan Z, Lu J, Li F, Qin F, Chang H (2018) Ethanol-mediated 2D growth of Cu_2O Nanoarchitectures on nanoporous Cu templates in anhydrous ethanol. Nano 2018(8):18. https://doi.org/10.3390/nano8010018

de Lima GF, Mavrandonakis A, de Abreu HA, Duarte HA, Heine T (2013) Mechanism of alcohol−water separation in metal−organic frameworks. J Phys Chem 117:24–30

de Oliveira JAR, da Silva Martins LH, Komesu A, Neto JM (2017) Nanotechnology applications on lignocellulosic biomass pretreatment. In: Rai M, da Silva S (eds) Nanotechnology for bioenergy and biofuel production, Green chemistry and sustainable technology. Springer, Cham, pp 19–37. https://doi.org/10.1007/978-3-319-45459-7

Devi A, Singh A, Bajar S, Owamah HI (2021) Nanomaterial in liquid biofuel production: applications and current status. Environ Sustain 4:343–353

Dong RJ, Zheng DF, Yang DJ, Qiu XQ (2019) pH-responsive lignin-based magnetic nanoparticles for recovery of cellulase. Bioresour Technol 294:122133

Fathy NA (2017) Carbon nanotubes synthesis using carbonization of pretreated rice straw through chemical vapor deposition of camphor. RSC Adv 7:35–41

Gupta K, Chundawat TS (2020) Zinc oxide nanoparticles synthesized using Fusarium oxysporum to enhance bioethanol production from rice-straw. Biomass Bioenergy 143:105840

Hamdi AMA, Sillanpaa M, Dutta J (2015) Photocatalytic degradation of phenol by iodine doped tin oxide nanoparticles under UV and sunlight irradiation. J Alloys Compd 618:66–71

Hamdi AMA, Sillanpaa M, Bora T, Dutta J (2016) Efficient photocatalytic degradation of phenol in aqueous solution by SnO_2:Sb nanoparticles. Appl Surf Sci 370:29–36

Huang PJ, Chang KL, Hsieh JF, Chen ST (2015) Catalysis of rice straw hydrolysis by the combination of immobilized cellulase from Aspergillus Niger on β-cyclodextrin-Fe_3O_4 nanoparticles and ionic liquid. Biomed Res Int 2015:409103

Hussein HT, Kareem MH, Hussein AMA (2021) Synthesis and characterization of carbon nanotube doped with zinc oxide nanoparticles CNTs-ZnO/PS as ethanol gas sensor. Optik 248:168107

Jorge EYC, Lima TDM, Lima CGS, Marchini L, Castelblanco WN, Rivera DG, Urquieta-Gonzalez EA, Varma RS, Paixao MW (2017) Metal-exchanged magnetic β-zeolites: valorization of lignocellulosic biomass-derived compounds to platform chemicals. Green Chem 19:3856–3868

Kang Q, Appels L, Tan T, Dewil R (2014) Bioethanol from lignocellulosic biomass: current findings determine research priorities. Sci World J 2014:298153. https://doi.org/10.1155/2014/298153

Kaur P, Taggar MS, Kalia A (2021) Characterization of magnetic nanoparticle–immobilized cellulases for enzymatic saccharification of rice straw. Biomass Conv Bioref 11:955–969

Kaur M, Gill MK, Sharma S, Kocher GS, Sodhi HS (2023) Biological pretreatment strategies for second-generation lignocellulosic biomass to enhance ethanol production. In: Agroindustrial waste for green fuel application. Springer Nature Singapore, Singapore, pp 169–203

Kim KY, Lee H (2016) Use of magnetic nanoparticles to enhance bioethanol production in syngas fermentation. Bioresour Technol 204:39–44

Kumar A, Singh S, Tiwari R, Goel R, Nain L (2017) Immobilization of indigenous holocellulase on iron oxide (Fe_2O_3) nanoparticles enhanced hydrolysis of alkali pretreated paddy straw. Int J Biol Macromol 96:38–49

Kushwaha D, Upadhyay S, Mishra P (2018) Nanotechnology in bioethanol/biobutanol production. Green Nanotechnol Biofuel Prod 6:115–127

Ladole MR, Mevada JS, Pandit AB (2017) Ultrasonic hyperactivation of cellulase immobilized on magnetic nanoparticles. Bioresour Technol 239:117–126

Lai HK, Chou YZ, Lee MH, Lin KYA (2018) Coordination polymer-derived cobalt nanoparticle-embedded carbon nanocomposite as a magnetic multi-functional catalyst for energy generation and biomass conversion. Chem Eng J 332:717–726

Lee SM, Jin LH, Kim JH, Han SO, Na HB, Hyeon TH, Koo YM, Lee JH (2010) B-Glucosidase coating on polymer nanofibers for improved cellulosic ethanol production. Bioprocess Biosyst Eng 33:141

Lee YC, Dutta S, Wu KC (2014) Integrated, cascading enzyme-/chemocatalytic cellulose conversion using catalysts based on mesoporous silica nanoparticles. ChemSusChem 12:3241–3246

Leo VV, Singh BP (2018) Prospectus of nanotechnology in bioethanol production. In: Srivastava N et al (eds) Green nanotechnology for biofuel production, biofuel and biorefinery technologies, vol 5. Springer, Cham, pp 129–139. https://doi.org/10.1007/978-3-319-75052-1_9

Li W, Ma S, Li Y, Yang G, Mao Y, Luo J, Gengzang D, Xu X, Yan S (2015) Enhanced ethanol sensing performance of hollow ZnO–SnO_2 core–shell nanofibers. Sensors Actuators B Chem 211:392–402

Li H, Han W, Lv R, Zhai A, Li XL, Gu W, Liu X (2019) Dual-function mixed-lanthanide metal–organic framework for ratiometric water detection in bioethanol and temperature sensing. Anal Chem 91:48–54

Li GB, Chen J, Qiao Song B, Zhang X, Zhang Z, Pan RK, Pei LM, Liao LS, Guan GW, Wang J, Liu SG, Yang QY (2022) Efficient purification of bioethanol by an ethanol-trapping coordination network. Sep Purif Technol 293:121097

Lin L, Liu T, Zhang Y, Liang X, Sun R, Zeng W, Wang Z (2016) Enhancing ethanol detection by heterostructural silver nanoparticles decorated polycrystalline zinc oxide nanosheets. Ceram Int 42:3138–3144

Madaci A, Suwannin P, Raffin G, Hangouet M, Martin M, Ferkous H, Bouzid A, Bausells J, Elaissari A, Errachid A, Jaffrezic-Renault N (2023) A sensitive micro conductometric ethanol sensor based on an alcohol dehydrogenase-gold nanoparticle chitosan composite. Nanomaterials 13(16):2316. https://doi.org/10.3390/nano13162316

Mahmoud HR (2014) Highly dispersed Cr_2O_3–ZrO_2 binary oxide nanomaterials as novel catalysts for ethanol conversion. J Mol Catal A Chem 392:16–22

Markandan K, Chai WS (2022) Perspectives on nanomaterials and nanotechnology for sustainable bioenergy generation. Materials 15:7769

Nayeri MD, Nikkhah H, Zilouei H et al (2023) Immobilization of cellulase on graphene oxide coated with $NiFe_2O_4$ and Fe_3O_4 for hydrolysis of rice straw. Cellulose 30:5549–5571

Nizami AS, Rehan M (2018) Towards nanotechnology-based biofuel industry. Biofuel Res J 5(2): 798–799

Obradovic MD, Stancic ZM, Lacnjevac UC, Radmilovic VV, Gavrilovic-Wohlmuther A, Radmilovic VR, Gojkovic SL (2016) Electrochemical oxidation of ethanol on palladium-nickel nanocatalyst in alkaline media. Appl Catal B Environ 189:110–118

Park HJ, Driscoll AJ, Johnson PA (2018) The development and evaluation of β-glucosidase immobilized magnetic nanoparticles as recoverable biocatalysts. Biochem Eng J 133:66–73

Pena L, Ikenberry M, Hohn KL, Wang D (2012) Acid functionalized nanoparticles for pretreatment of wheat straw. J Biomater Nanobiotechnol 3:342–352

Rai M, dos Santos JC, Soler MF, Marcelino PRF, Brumano LP, Ingle AP, Gaikwad S, Gade A, da Silva SS (2016) Strategic role of nanotechnology for production of bioethanol and biodiesel. Nanotechnol Rev 5(2):231–250

Rai M, Ingle AP, Pandit R, Paralikar P, Biswas JK, da Silva SS (2019) Emerging role of nanobiocatalysts in hydrolysis of lignocellulosic biomass leading to sustainable bioethanol production. Catal Rev 61(1):1–26

Rana C, Bera SR, Saha S (2019) Growth of SnS nanoparticles and its ability as ethanol gas sensor. J Mater Sci Mater Electron 30:16–29

Razack SA, Duraiarasan S, Mani V (2016) Biosynthesis of silver nanoparticle and its application in cell wall disruption to release carbohydrate and lipid from C. vulgaris for biofuel production. Biotechnol Rep 11:70–76

Salehi N, Mirjalili BBF (2017) Synthesis of highly substituted dihydro-2-oxopyrroles using Fe_3O_4@nano-cellulose–OPO_3H as a novel bio-based magnetic nanocatalyst. RSC Adv 7: 30303–30309

Sanusi IA, Suinyuy TN, Kana GEB (2021) Impact of nanoparticle inclusion on bioethanol production process kinetic and inhibitor profile. Biotechnol Rep 29:e00585

Sheikh ZUD, Bajar S, Devi A, Rose PK, Suhag M, Yadav A, Yadav DK, Deswal T, Kaur J, Kothari R, Pathania D, Rani N, Singh A (2023) Nanotechnology based technological development in biofuel production: current status and future prospects. Enzym Microb Technol 171: 110304

Shibuya M, Sasaki K, Tanaka Y, Yasukawa M, Takahashi T, Kondo A, Matsuyama H (2017) Development of combined nanofiltration and forward osmosis process for production of ethanol from pretreated rice straw. Bioresour Technol 235:405–410

Singh N, Dhanya BS, Verma ML (2020) Nano-immobilized biocatalysts and their potential biotechnological applications in bioenergy production. Mater Sci Energy Technol 3:808–824

Sirajunnisa AR, Surendhiran D (2016) Algae—a quintessential and positive resource of bioethanol production: a comprehensive review. Renew Sust Energ Rev 66:248–267

Tiempos-Flores N, Hernandez-Fernández E, Rico-Barragan A, Alvarez JR, Juárez-Ramírez I, Garza-Navarro MA, Rodríguez-Hernández J, Fonseca-García A, Michaelis DJ, Davila-Guzman NE (2022) Enhanced hydrophobicity of modified ZIF-71 metal-organic framework for biofuel purification. Polyhedron 217:115736

Trisunaryanti W, Triyono, Fatmawati DA (2020) Synthesis of Co-NH_2/mesoporous silica bifunctional catalyst using sidoarjo mud and bovine bone gelatin template for conversion of used cooking oil into biofuel. Rasayan J Chem 13:723–732

Tropea A (2022) Biofuels production and processing technology. Fermentation 8:319

Verma ML, Barrow CJ, Puri M (2013) Nanobiotechnology as a novel paradigm for enzyme immobilization and stabilization with potential applications in biodiesel production. Appl Microbiol Biotechnol 97:23–39

Wang W, Ji S, Lee I (2013) Fast and efficient nanoshear hybrid alkaline pretreatment of corn stover for biofuel and materials production. Biomass Bioenergy 51:35–42

Wang TT, Ma SY, Cheng L, Luo J, Jiang XH, Jin WX (2015) Preparation of Yb-doped SnO_2 hollow nanofibers with an enhanced ethanol–gas sensing performance by electrospinning. Sensors Actuators B Chem 216:212–220

Wang DJ, Huang DR, Feng Z, Liu H, Su D (2016a) Multi-walled carbon nanotubes as a catalyst for gas-phase oxidation of ethanol to acetaldehyde. ChemSusChem 9:1820–1826

Wang Y, Liu L, Meng C, Zhou Y, Gao Z, Li X, Cao X, Xu L, Zhu W (2016b) A novel ethanol gas sensor based on $TiO_2/Ag_{0.35}V_2O_5$ branched nanoheterostructures. Sci Rep 6:33092

Wang Y, Li H, Huang D, Wang X, Cai L, Chen Y, Wang W, Song Y, Han G, Zhen B (2022a) A high-performance ethanol gas sensor based on Ce-doped SnO_2 nanomaterials prepared by the Pechini method. Mater Sci Semicond Process 137:106188

Wang Z, Yang C, Li X, Song X, Pei C, Zhao ZJ, Gong J (2022b) The role of CO_2 dissociation in CO_2 hydrogenation to ethanol on CoCu/silica catalysts. Nano Res 16:28–33
Wu Y, Li X, Liu JH, He YN, Yu LM, Liu WH (2012) ZnO nanomaterials based surface acoustic wave ethanol gas sensor. J Nanosci Nanotechnol 12:5–9
Wu J, Gao G, Li J, Sun P, Long X, Li F (2017) Efficient and versatile CuNi alloy nanocatalysts for the highly selective hydrogenation of furfural. Appl Catal B Environ 203:27–36
Zhong R, Sels BF (2018) Sulfonated mesoporous carbon and silica-carbon nanocomposites for biomass conversion. Appl Catal B Environ 236:518–545

Chapter 7
Fabrication of Activated Charcoal from Paddy Waste for Bioethanol Production

Gyanendra Tripathi, Diksha Sharma, Parnika Mishra, Vaishnavi Mishra, Irum, Priyanka Dubey, Suhail Ahmad, Vishal Mishra, and Alvina Farooqui

Abstract The global pursuit of sustainable energy sources has led to the emergence of bioethanol as a promising alternative to fossil fuels. This renewable biofuel is derived from biomass feedstocks through fermentation and holds the potential to mitigate greenhouse gas emissions and enhance energy security. However, the efficient production and purification of bioethanol remain pivotal challenges. Traditional purification methods often fall short in removing impurities and water effectively, necessitating innovative strategies. Activated charcoal, known for its remarkable adsorption properties, offers a sustainable solution for bioethanol purification. In this paper we explored the fabrication of activated charcoal for bioethanol production, examining its characteristics, preparation methods, and applications in ethanol purification. By analysing existing literature, we shed light on the advances, limitations, and implications of incorporating activated charcoal into the bioethanol production process. This chapter will contribute to the understanding of how activated charcoal can enhance the efficiency and sustainability of bioethanol purification, fostering the progression towards a cleaner and more renewable energy future.

Keywords Bioethanol · Activated charcoal · Carbonization · Pretreatment · Paddy straw

G. Tripathi · S. Ahmad · A. Farooqui (✉)
Department of Bioengineering, Integral University, Lucknow, Uttar Pradesh, India
e-mail: alvina@iul.ac.in

D. Sharma
Institute of Biomedical and Research, Mangalaytan University, Aligarh, Uttar Pradesh, India

P. Mishra · V. Mishra · Irum · P. Dubey
Department of Biosciences, Integral University, Lucknow, Uttar Pradesh, India

V. Mishra (✉)
School of Biochemical Engineering, IIT (BHU), Varanasi, Uttar Pradesh, India
e-mail: Vishal.bce@itbhu.ac.in

N. Srivastava et al. (eds.), *Paddy Straw Waste for Biorefinery Applications*, Clean Energy Production Technologies, https://doi.org/10.1007/978-981-99-8224-0_7

7.1 Introduction

In the face of global energy and environmental challenges, the pursuit of sustainable and renewable energy sources has become imperative. One of the promising contenders in this arena is bioethanol—a biofuel derived from biomass feedstocks through the process of fermentation. Bioethanol possesses the potential to significantly reduce our reliance on fossil fuels, mitigate greenhouse gas emissions, and enhance energy security. However, the efficient production and purification of bioethanol remain key challenges for realizing its widespread adoption (Farooqui et al. 2021). In this context, activated charcoal, known for its exceptional adsorption properties, has emerged as a promising solution for the purification of bioethanol. This paper delves into the intricacies of fabricating activated charcoal for bioethanol production, examining its characteristics, preparation methods, and applications in ethanol purification. By critically assessing existing literature, we aim to illuminate the advances, limitations, and implications of incorporating activated charcoal in the bioethanol production process (Chen et al. 2021).

Bioethanol, an alcohol primarily composed of ethanol, is generated through the microbial fermentation of various biomass feedstocks, including agricultural residues, lignocellulosic materials, and sugarcane molasses. The fermentation process yields a crude ethanol-water mixture, laden with impurities such as residual sugars, organic acids, and biomass remnants. The subsequent purification step is essential to achieve high ethanol concentrations suitable for utilization as a transportation fuel. Traditional methods of bioethanol purification, such as distillation and molecular sieves, have limitations in terms of energy consumption, cost, and efficiency. These methods often struggle to remove water and impurities effectively, leading to increased energy expenditures during distillation and product losses due to entrainment of valuable ethanol. As the demand for higher bioethanol purity intensifies, there arises a need for innovative and sustainable purification strategies that address these challenges.

Activated charcoal, a porous carbon material with an extensive surface area, has long been recognized for its exceptional adsorption properties. The activation process involves the controlled oxidation of carbonaceous precursors, creating a network of pores and surface sites capable of adsorbing a diverse range of molecules. This unique adsorption behaviour stems from the combined effect of physical and chemical interactions, including van der Waals forces, electrostatic attractions, and π-π interactions (Abd-Rabboh et al. 2022).

The application of activated charcoal extends across diverse industries, encompassing waste water treatment, air purification, and pharmaceuticals. In the context of bioethanol production, activated charcoal's remarkable adsorption capacity, coupled with its ability to selectively adsorb impurities and colorants, positions it as an attractive candidate for ethanol purification. The regenerability of activated charcoal further enhances its appeal, offering a sustainable solution that aligns with the principles of a circular economy.

The production of activated charcoal involves a sequence of steps, including precursor selection, carbonization, and activation. Carbonaceous precursors, which can range from agricultural residues to wood-based materials, undergo controlled thermal treatment under inert conditions to form carbon-rich intermediates. Subsequent activation, achieved through physical or chemical methods, creates a porous structure characterized by micropores, mesopores, and macropores. The specific surface area, pore size distribution, and surface chemistry of the activated charcoal can be tailored to suit the intended application. In the context of ethanol purification, the design and fabrication of activated charcoal involve optimizing parameters such as precursor type, activation temperature, and activation time. The choice of precursor significantly influences the properties of the activated charcoal, with variations in elemental composition, pore structure, and adsorption characteristics. Furthermore, the activation process type, whether steam activation or chemical activation, dictates the final adsorption performance. These variables collectively influence the activated charcoal's affinity for water, impurities, and colorants present in the bioethanol-water mixture (Singh et al. 2022; Roy 2023). In this chapter, we have explored the fabrication of activated charcoal for bioethanol production, focusing on its properties, preparation methods, and potential applications in ethanol purification. By delving into the scientific literature, we aim to provide insights into the advancements and challenges associated with utilizing activated charcoal as an adsorbent in the bioethanol industry. Furthermore, the environmental and economic implications of integrating activated charcoal into the bioethanol production process will be discussed.

7.2 Paddy Straw

Rice holds a prominent status among India's primary food crops, both in terms of the land dedicated to its cultivation and the overall yield. India ranks as the second-largest global producer of rice, trailing only behind China. Owing to this, it also produces a large amount of waste or by-product. The process of cultivating rice results in two distinct by-products, i.e. rice straw and rice husk. In India, there is a substantial abundance of rice straw, making it readily accessible for the production of bioethanol (Singh et al. 2016).

Paddy straw stands out as a significant crop residue, resulting from the harvesting of rice crops. The improper disposal of paddy straw, including unregulated decomposition and the harmful practice of stubble burning, poses serious environmental concerns. These practices not only endanger the health of living organisms but also contribute to climate change by releasing substantial amounts of greenhouse gases. It holds a promise as a valuable resource due to its rich content of hexose (C_6) and pentose (C_5) sugars. This abundance of sugars makes paddy straw a potentially valuable source for ethanol production. This involves breaking down the polysaccharides in paddy straw into simpler sugars through hydrolysis, followed by fermentation to produce ethanol (Hans et al. 2019). Rice straw can be described as a fibrous substance primarily composed of carbohydrate polymers, such as cellulose and hemicellulose. It is intricately bound, either through covalent or non-covalent bonds, with lignin. Additionally, rice straw contains various extractive substances and inorganic compounds (Singh and Kumar 2019).

7.2.1 Components of Paddy Straw

Among agricultural crop residues, rice straw boasts the highest cellulose content worldwide. The predominant constituents of rice straw include cellulose (making up 33–47% of its composition), hemicelluloses (comprising 19–27%), lignin (ranging from 5–24%), ash (comprising 13–20%), and a notable presence of silica (approximately 18.3%). This high cellulose content makes rice straw an exceptionally abundant source of cellulose in the agricultural residue category (Singh et al. 2016) (Fig. 7.1).

7.2.1.1 Lignin

For the creation of useful chemicals, preserving carbon balance, and manufacturing energy and fuels, it is essential to fragment the complex lignin structure into smaller monomeric molecules using cutting-edge methods (Pulicharla et al. 2016). The creation of lower molecular weight substances including vanillin, eugenol, lignosulfonates, hydroxylated aromatics, aldehydes, aliphatic acids, and others is possible thanks to the thermochemical depolymerization of lignin. Among these techniques, base-catalysed depolymerization (BCD) is the most widely used and extensively studied method for obtaining phenolic monomers by cleaving the predominant (60%) β-O-4 bond in lignin (Kumari et al. 2008). Besides, other organic acids, including citric acid, lactic acid, and acetic acid, have also been produced from lignin-rich wastes. In the food processing sector, these organic acids are used as acidulants or preservatives to protect food against microbial degradation (Yanti et al. 2012). The leather industry uses lignosulfonate, a lignin derivative that is insoluble in water, as a tanning agent. In a particular study, lignosulfonate obtained from rice straw was employed for leather tanning (Nasr et al. 2017). Moreover, as demonstrated in a study, the sulfonation of lignin enhances its effectiveness, enabling its

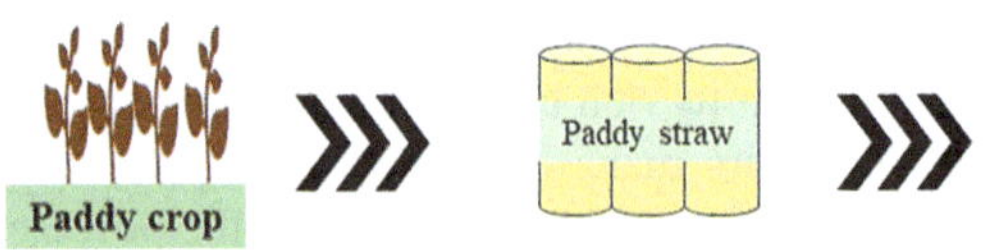

Cellulose	33-47%
Hemicellulose	19-27%
Lignin	5-24%
Ash	13-20%
Silica	18.3%

Fig. 7.1 Components of paddy straw

usage as a binding material in the cement and dyeing industries (Aro and Fatehi 2017).

7.2.1.2 Cellulose and Hemicellulose

Rice straw encompasses C_6 and C_5 sugars among its constituents, which can be subjected to microbial fermentation to yield desired biochemicals. Hemicellulose, for instance, can be disintegrated into its individual components like arabinose, xylose, mannose, galactose, ribose, lyxose, and more. These components, characterized by an undefined structure, can be harnessed to produce all useful chemical derivatives such as Xylitol, 5-aminovalaric acid, 1,5-diaminopentane, itaconic acid, and levulinic acid and other compounds, as extensively detailed in studies focusing on hemicellulose (Kumari et al. 2008). Notably, this process diverges from starch primarily in the arrangement and bonding of glucose units.

Conversely, cellulose can be disintegrated into its individual monomeric components through the enzymatic action of microorganisms. The fermentation of these intricate sugars has been observed to generate a diverse array of organic acids, including acetic acid, ferulic acid, lactic acid, formic acid, succinic acid, malic acid, levulinic acid, aspartic acid, and numerous others. Extensive documentation exists regarding the potential health advantages of these acids, including their anti-diabetic, anti-cancer, neuroprotective, and antioxidant properties (Norrahim et al. 2021; Sun et al. 2019).

7.2.1.3 Ash

Ash is another substantial by-product that results from the pyrolysis of waste biomass, in addition to lignin, cellulose, and hemicellulose. Based on variables including harvest time, climatic conditions, and the location of the production site, the ash content may vary. Ash concentration in rice straw typically ranges from 13% to 20% (El-Sayed and El-Samni 2006). As rice straw is transformed in biorefineries to create products with value-added goods, partially burnt residues in the form of ash can persist. Despite being commonly seen as waste, previous studies have highlighted that rice straw ash contains a notable amount of silica, which is a chemically and thermally stable composite. This silica can be effectively utilized as a binding material (Do et al. 2020). In older times, silica was collected from the waste produced during the processing of rice straw, and its significance as a significant inorganic material has been acknowledged in a number of applications, such as bioremediation, medicinal treatments, and the cement industry (Nandiyanto et al. 2016).

7.2.1.4 Silica

Silica is a crucial component with diverse applications, including biotechnology materials (Galliker et al. 2010; Cheang et al. 2012), environmental treatment (Liu et al. 2013, cement production (Singh et al. 2011), composite materials (Nandiyanto et al. 2014; Nandiyanto et al. 2011), and medical applications (Kwon et al. 2013; Estévez et al. 2009; Lu et al. 2007). Its popularity stems from its outstanding properties such as chemical, physical, and thermal stability, compatibility with various materials, wide availability, and cost-effectiveness (Galliker et al. 2010; Cheang et al. 2012). Rice plant-related waste presents a promising source of silica, particularly in tropical regions like Indonesia. Utilizing rice straw as a silica source offers not only economic benefits but also environmental advantages (Lu and Hsieh 2012).

7.2.2 Applications of Paddy Straw in Various Sectors

Paddy straw, an abundant agricultural residue generated from rice cultivation, has gained significant attention due to its potential applications across various sectors. Historically considered a waste material and often burned after harvesting, paddy straw is now recognized as a valuable resource with diverse uses. This versatile residue can be harnessed in sectors ranging from agriculture and energy to construction and environmental conservation. In this paper, we explore the multifaceted applications of paddy straw, highlighting its contributions to sustainability, waste reduction, and economic growth. By investigating the utilization of paddy straw in different sectors, we aim to underscore its role in creating a more circular and resource-efficient agricultural landscape. Figure 7.2 represents the role of paddy straw in various fields which are briefly explained below.

7.2.2.1 Production of Biogas from Paddy Straw

Producing biogas from rice straw entails placing shredded straw within an anaerobic digester, where microorganisms decompose its organic elements. These results in generating biogas with blend of methane and carbon dioxide. This could be serving as a valuable source of energy for various applications, including power generation, cooking, and heating. The by-product of this process, known as digestate, is rich in nutrients and can be effectively utilized as a fertilizer. This approach provides advantages such as waste reduction, renewable energy production, and the recycling of nutrients. However, it necessitates careful management of the digester to achieve optimal outcomes (Singh and Brar 2021).

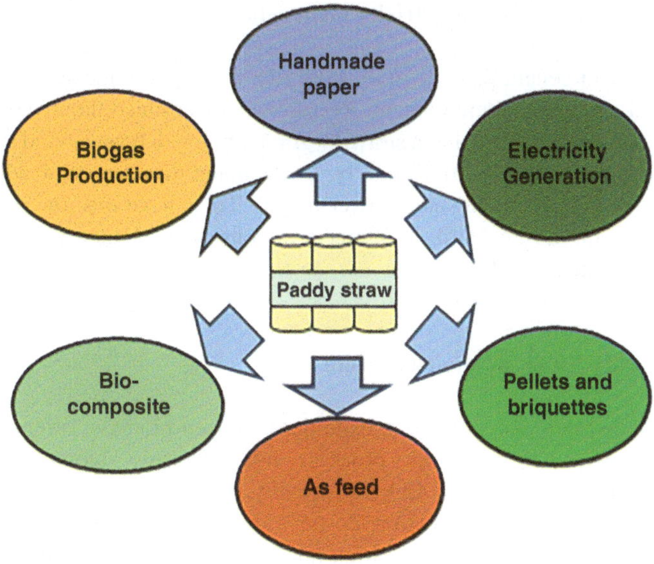

Fig. 7.2 Applications of paddy straw in various fields

7.2.2.2 Paddy Straw as Hand-Made Paper

The production of biogas from rice straw involves placing shredded straw into an anaerobic digester, where microorganisms break down its organic components. This process yields biogas, a mixture of methane and carbon dioxide, which serves as a valuable energy source for applications like power generation, cooking, and heating. The resulting by-product, called digestate, is nutrient-rich and can be used as an effective fertilizer. This method offers benefits like waste reduction, renewable energy generation, and nutrient recycling, but it requires meticulous digester management to ensure the best results (Pushpa and Sinha 2011).

7.2.2.3 Paddy Straw as Feed

In India, rice straw has been a traditional feed for cattle and buffaloes for generations. In some regions, there may be a preference for feeding cattle and buffaloes a combination of green fodder and whole wheat straw due to the presence of "Degnala" disease, which is a prevalent issue. In areas where green fodder is in short supply, rice straw is commonly used as the primary diet at home. Additionally, after rice harvesting, livestock often graze on the remaining stubbles and ratoon in the fields (Singh et al. 1995).

7.2.2.4 Paddy Straw for Electricity Generation

The key factor in generating electricity from rice straw lies in the availability of the raw material and its moisture content. Researchers determined the maximum availability of raw materials by using a straw-to-grain ratio, which was found to be nearly 0.75 (Logeswaran et al. 2020). Similarly, it was noted that rice straw with reduced moisture content leads to elevated power output. To achieve this, the biomass was subjected to sun-drying until it attained a consistent weight prior to further processing (Gupta et al. 2015).

7.2.2.5 Paddy Straw as Bio-composite

Strawboard, a widely used bio-composite derived from biomass, offers a mutually beneficial opportunity. In this process, lignocellulosic biomass undergoes pretreatment with supercritical CO_2, effectively elevating the fibre's value while simultaneously reducing production costs. Moreover, this biomass serves as a valuable resource for crafting high-quality straw powder/polylactic acid bio-composites (Yu et al. 2020).

7.2.2.6 Paddy Straw as Pellets and Briquettes

The process of producing pellets and briquettes involves several steps, including cutting, drying, compacting, and binding loose straw using adhesives (Purohit and Chaturvedi 2018; Nolan et al. 2010). In the process of pelletizing biomass, the finely crushed biomass is compacted to enhance its density, resulting in the formation of smaller-sized pellets. These (pellets) serve as a substitute for traditional coal, known as bio-coal, which offers advantages such as ease of storage and transportation, as well as high efficiency when compared to fossilized coal (Singh and Brar 2021).

7.3 Steps Involved in the Transformation of Rice Husk to Activated Biochar

Due to its substantial surface area and excellent adsorptive capacity, AC has the potential to be used as a carrier in solid forms. Since its application in the treatment of industrial wastewater and gases to achieve environmental protection and material recovery objectives, AC consumption has expanded globally (Naji and Tye 2022). Utilizing economical resources for the preparation of AC is crucial. This involves using low-cost agricultural and forest residues, considering the substantial expenses associated with both AC production and regeneration processes. These carbonaceous materials' great abilities to maintain the physiochemical attributes and

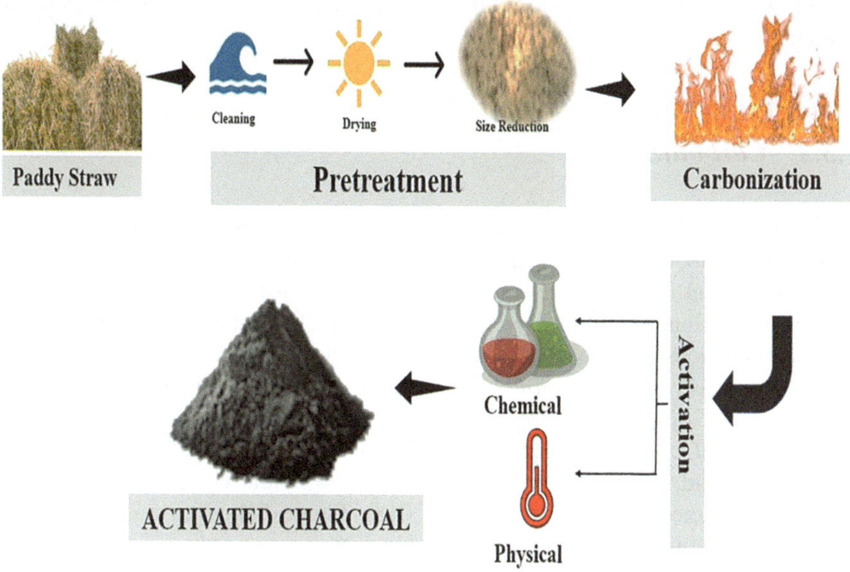

Fig. 7.3 Steps involve in activated charcoal production from paddy straw

minimize inorganic constituents, hardness, low ash content, and elevated content of carbon have all demonstrated that they are effective precursors to the production of AC. Therefore, to make AC, any material with a high carbon concentration and little inorganic component can be used (Narowska et al. 2020). Rice husks having carbon content of 20.988%, volatiles content of 68.08%, ash content of 0.952%, and moisture content of 10% can be used as a precursor for activated carbon. Figure 7.3 represents the steps involved in activated charcoal production from paddy straw and is explained as follows.

7.3.1 Pretreatment

Prior to activation, these carbon-based materials undergo a series of preparatory steps converting them into AC. These steps include (i) cleansing to eliminate contaminants (Tripathi et al. 2022), (ii) dehydration to eliminate any residual moisture that might impact the subsequent procedures (Sadeek et al. 2020), (iii) grinding and sifting to acquire the carbon source materials (Mazlan et al. 2016), and (iv) removing mineral content using acidic or alkaline solutions (like HCl and hydrofluoric acids). The minimal amount of ash content and minerals are required for AC to function as a catalyst (Phiri et al. 2017). Gao et al. (2011) suggested that before cutting raw rice straw into 2–3 cm lengths, soaked it in NaOH solution (2 wt. %) for 2 days to eliminate ash and other water-miscible materials. After reaching neutrality, the rice straw was rinsed in distilled water, dried, and placed in a storage container.

7.3.2 *Carbonization*

Biomass thermochemical conversion employs heat to start the decomposition process in order to transform organic materials into a more useful energy source. Thermochemical conversion methods encompass co-firing, gasification, liquefaction, carbonization, and pyrolysis as outlined by Tamalouzt et al. (2014). As all these chemical reactions leading to the formation of solid "char", liquid "bio-oil", and gaseous "biogas" occur within environments devoid of oxygen or with restricted oxygen levels, pyrolysis is considered the fundamental principle underlying all thermochemical conversion techniques. The volatile component of biomass is where gases and bio-oil are produced, as opposed to the char, which is mostly composed of fixed carbon (Sahri et al. 2018). Figure 7.4 gives a diagrammatic view and elaboration of processes involved in carbonization step.

Carbonization is a process that turns organic material into elemental carbon by removing non-carbon (volatile components). In this process, any material that contains noticeable carbon is pyrolysed at temperatures between 600 and 1000 °C, frequently within an unreactive atmosphere containing gases such as nitrogen (N_2) or argon. Precursors may have their physicochemical features retained by being heated over 800 °C during the carbonization stage, producing AC with a high thermal stability (Naji and Tye 2022).

The different process regions and heating stages of carbonization are explained as follows:

- During the pre-drying phase of the carbonization process, the biomass is gradually heated from ambient temperature to approximately 100 °C, which serves as the drying temperature.
- **Drying:** The only purpose of the heat used at this point is to move the biomass's moisture content from the liquid to the vapour phase. Due to the steady temperature of evaporation during this stage, the temperature of the biomass remains nearly constant regardless of the exterior temperature of the reactor. The raw biomass should be sun-dried before being added to the reactor in order to minimize heat loss during this phase.
- **Post drying:** The temperature at this stage varies from 100 to 200 °C. The biomass doesn't begin to decompose significantly at this point. During this

Fig. 7.4 Process involve in carbonization step

step, only the water molecules locked inside the fibres and very light volatiles are released. This process is typically quick and doesn't require a lot of heat.

- **Torrefaction** takes place between temperatures of 200 and 300 °C. The breakdown twitches at this point. Hemicellulose is the first component of the biomass to begin breaking down. Nearly all of the hemicellulose has been broken down at this point, and minor amounts of cellulose have also begun to break down. Torrefaction produces a product with a high energy density and low fixed carbon content if the heating process is paused at this point to maintain the temperature.
- **Low-temperature carbonization** takes place at temperatures between 300 and 400 °C. Hemicellulosic and cellulose breakdown entirely at this stage, and lignin begins to deteriorate. In comparison to torrefaction, the fixed carbon content produced by the breakdown and decomposition of the biomass is marginally higher. Lower energy density is the outcome, though.
- **High-temperature carbonization**: Charcoal that is extremely high in carbon is produced if heating is allowed to go past 400 °C. The final temperature affects how much lignin is degraded. Very high fixed carbon levels are obtained at this stage. Rao et al. (2022) in their study maintained the carbonization temperature at 500 °C for 100 min while rising at a rate of 10 °C/min. Husk and straw yields for carbonization were 44% and 37%, respectively. After that, a vessel was attached to the muffle furnace's base to provide steam to the machinery, keeping the mass ratio of water to carbonizate at 2:1.

7.3.3 Activation

By means of activation procedures, the porous framework of the AC established during carbonization can be more extensively developed. These activation procedures work to increase the volume of pores, expand the size of pores, and raise the overall surface area of the activated carbon. Improving the microporous arrangement is attainable by removing disordered carbon and efficiently integrating lignin into the activation process alongside the activating agent. The incineration of the walls separating the pores often leads to their expansion. Consequently, following this stage, there is noticeable growth in both temporary pores and microporosity. Macropores >25 nm, 1 nm < mesopores <25 nm, and micropores <1 nm were the three classification systems employed to categorize the pores developed on the surface of AC (Naji and Tye 2022). There are two methods for doing the activation step: chemical and physical.

7.3.3.1 Physical Activation

Carbonaceous materials are usually physically activated in two phases. After the carbonization procedure, the resulting char is subjected to activation using oxidizing gases like CO_2, steam, or a mixture of air, at elevated temperatures ranging from

800 to 1000 °C. This process leads to the formation of a porous structure characterized by a substantial surface area, brought about by the reactions described below (7.1) to (7.3) (Mazlan et al. 2016).

$$C + O_2 \rightarrow CO_2 \tag{7.1}$$

$$C + CO_2 \rightarrow 2CO \tag{7.2}$$

$$C + H_2O \rightarrow CO_2 + 2H_2 \tag{7.3}$$

In comparison to chemical activation, physical activation has a number of benefits, including a lower activation cost and the absence of chemical wastes (Pallarés et al. 2018). The primary downsides of physical activation, on the other hand, are its prolonged activation time and high energy consumption as a result of its twin cooling phases (Yahya et al. 2015). The AC generated through this method also exhibits certain deficiencies due to the necessity of using high temperatures (up to 1000 °C) during the activation phase. Consequently, it is not suitable for applications requiring catalyst, adsorbent, or filtration functions (Tadda et al. 2016). Steam and carbon dioxide are the primary and most valuable activation agents in physical activation. Due to their smaller size of 0.265 nm, H_2O molecules are more accessible to carbon pores than CO_2 molecules as suggested in the study by Abd et al. (2021). As a result, the reaction will happen quickly, creating new pores. Another benefit of employing steam as an activator is that the AC produced has a lot of rapid increase in surface area within a brief reaction period. Yang et al. (2010) conducted a study on utilizing CO_2, steam, and a combination of both for physically activating coconut shell-derived activated carbon. The findings revealed that the duration for CO_2 activation was approximately 2.5 times longer compared to the time required for steam activation or a combination of steam and CO_2. This disparity in time led to the slowest reduction rate in AC yield for CO_2 activation, in contrast to steam activation. Nonetheless, this heightened reactivity between carbon and the activating agent, while demanding longer activation periods, brought about an elevated BET extent of surface and volume of pores. Consequently, the augmentation in surface area and pore volume contributed to the enhancement and formation of fresh pores. Thus, CO_2 activation yielded a greater increase in AC surface area compared to both steam activation and the steam-CO_2 activation combination. The extent of pore creation exhibited a notable rise with increasing activation time for all three distinct activating agents. The advantages of CO_2 activation stem from its purity and uncomplicated management of gas flow. Additionally, compared to groups produced by steam activation, a series of groups formed by CO_2 activation, such as CO, are more stable (Pallarés et al. 2018).

Pallarés et al. (2018) investigated the effectiveness of CO_2 and steam as activation agents for producing activated carbon from barley straw, finding a similar trend. The outcomes indicated that the use of CO_2 activation significantly heightened the microporosity of the resulting AC, yielding impressive BET surface area and micropore volume of 789 m^2/g and 0.3268 cm^3/g, respectively. In contrast, the AC produced via steam activation displayed values of 552 m^2/g and 0.2304 cm^3/g

for surface area measured by BET analysis, and the volume of micropores exhibited different levels of microporosity development. This distinction can be attributed to the increased reactivity of steam at higher temperatures, leading to the formation of larger and wider pores. As a consequence, this promotes the enlargement of mesoporosity over microporosity. As a result of its high capacity to produce ACs with a small micropore structure and delayed reaction time, physical activation using carbon dioxide is strongly recommended (Tadda et al. 2016).

Based on the weight of the husk and straw used, Rao et al. performed the activation at 800 °C, and the amount of AC produced varied from 27% to 29% (Rao et al. 2022). Utilizing a batch-type pyrolysis reactor, Sakhiya et al. (2022) generated biochar from rice straw through slow pyrolysis in the temperature range of 300 to 500 °C. This process involved a residence time of 60 min and a heating rate of 10 °C/min. To guarantee the reactor was in an inert state, the air was first removed from it using a vacuum pump, and then 1.5 L/min of nitrogen was used to purge the reactor (vessel). The produced biochar cooled at room temperature after 1 min of operation in the batch-type pyrolysis reactor.

7.3.3.2 Chemical Activation

Wet oxidation is a common one-step method for chemically activating carbonaceous materials. This method is frequently used for feedstocks abundant in cellulose content, as noted by Yahya et al. (2015). The process involves the impregnation or blending of carbonaceous materials with oxidants and dehydrating agents (such as KOH, $ZnCl_2$, H_3PO_4, K_2CO_3, etc.) for chemical activation. Subsequently, the impregnated mixture is subjected to simultaneous activation and carbonization accomplished in a solitary phase under moderate conditions ranging from 400 to 600 °C temperatures. This procedure results in the production of AC possessing appropriate porosity, in accordance with findings by Tadda et al. (2016). This chemical activation technique offers advantages such as minimal energy requirements, short activation durations, substantial microporosity on surfaces, and achieving surface areas of up to 3000 m^2/g (Abd et al. 2020). The main disadvantage of the chemical activation process, however, is the contaminated wastewater that results from the chemical washing. The effectiveness of the chemical activation process relies significantly on various factors. These elements include the choice of the activating agent, the temperature and duration of activation, the ratio of activating agent to dry carbonaceous material, the method of impregnation, and the incorporation of nitrogen or argon flow during carbonization (Yorgun et al. 2016). The choice of activation agent is a crucial factor in creating an activated surface with exceptional textural characteristics. Since the evident change in the thermal breakdown of precursors, these agents reduce the creation of bitumen, resulting in an elevated yield of AC. Additionally, activating chemicals engage in redox reactions with the carbon surface to etch minute pores into the structure and enhance surface area (Heidarinejad et al. 2020).

Typical substances have the potential to serve as agents (KOH, $ZnCl_2$, NaOH, $AlCl_3$, H_3PO_4, K_2CO_3, and $MgCl_2$) (Abd et al. 2020). According to Yahya et al. (2015), the employment of metal hydroxides, such as potassium hydroxide and sodium hydroxide, during the carbon activation procedure yielded higher adsorption capacity compared to other activating agents. Additionally, H_3PO_4 and $ZnCl_2$ are utilized to activate materials before carbonization, but they are also applied in activating the precursor substances for charcoal and char. As a result, phosphoric acid-based activation generally occurs between temperatures of 450 and 600 °C, whereas activation involving potassium hydroxide takes place within the range of 450 to 850 °C. Notably, phosphoric acid necessitates a lower activation temperature compared to potassium hydroxide. In contrast, sodium hydroxide (400–900 °C), potassium carbonate (700–1000 °C), and zinc chloride (550–850 °C) activation at respective temperatures. Yet, potassium hydroxide outperformed sodium hydroxide in terms of performance and morphology for activation in a number of applications (Byamba-Ochir et al. 2016). In comparison to KOH, NaOH is thought to be the agents that are cost-effective and have a reduced environmental impact (Byamba-Ochir et al. 2016). Additionally, too much activating agent reduces porosity and generates a layer with insulating properties covering the surface of the AC (Zhong et al. 2012). Nonetheless, its downside involves the presence of residual chemical agents on the char, necessitating an additional washing step for complete removal. Gao et al. (2011) conducted activation using $(NH_4)_2HPO_4$, which exhibited characteristics akin to H_3PO_4. They found that the resulting activated carbon had a surface area of 1350 m^2/g, with a predominantly microporous carbon structure. Meanwhile, Sharath et al. (2017) employed sodium hydroxide (NaOH) as the activating agent for chemically activating the activated carbon derived from rice husk pyrolysis. The characterization of chemically activated carbons was conducted through assessment of yield percentages and bulk densities. The resulting activated carbon yields from rice husk at various activation temperatures—650, 700, and 800 °C—were approximately 48% (with a bulk density of 0.2 g/mL), 47% (with a bulk density of 0.16 g/mL), and 46% (with a bulk density of 0.117 g/mL), respectively.

7.4 Role of Activated Charcoal in Bioethanol Production

Activated charcoal, also known as activated carbon, plays a significant role in bioethanol production, by effectively adsorbing impurities, facilitating dehydration, and enhancing the quality and purity of the final product. The roles of activated charcoal are given in Fig. 7.4 and are explained as follows.

7.4.1 Removal of Inhibitors

With several applications in purification, AC is a potential solid carrier. For the purpose of detoxification, activated carbon serves as an inhibitor adsorbent. The growth of biomass was influenced favourably by this detoxifying technique. It had a positive influence on the growth of biomass. There are a number of benefits of using AC, including its structure with pores, substantial adsorption ability, extensive surface area ranging from 500 to 5000 m^2/g/L, fully developed micropores, and diverse array of surface functional groups (Soleimani et al. 2015; Gupta et al. 2017).

A useful adsorbent in the synthesis of bioethanol is activated carbon, which has a porous structure and a large surface area. These inhibitors are drawn to and bound by the surface of activated carbon, which is rich in adsorption sites, using mild forces to pull them out of the fermentation medium. As a result, inhibitor concentrations are lower, which lessens their negative impact on yeast metabolism. By improving yeast viability, growth, and ethanol production, the whole bioethanol output is optimized (Fig. 7.5).

Due to its great sorption capacity and practical commercial use, activated carbon adsorption is of interest. When separating the hydrolysates, valuable hydrolysates such monosaccharide and soluble oligosaccharide should not be eliminated; hence selective removal of inhibitors is extremely desirable. These sugars serve as the starting point for the fermentation that produces ethanol (Duarte et al. 2005). In the findings of Klasson et al. (2013), the adsorption capacity of biochar helped lignocellulosic biofuel fermentation by eliminating inhibitory substances including furfural and 5-hydroxymethylfurfural (HMF) produced during biomass pretreatment.

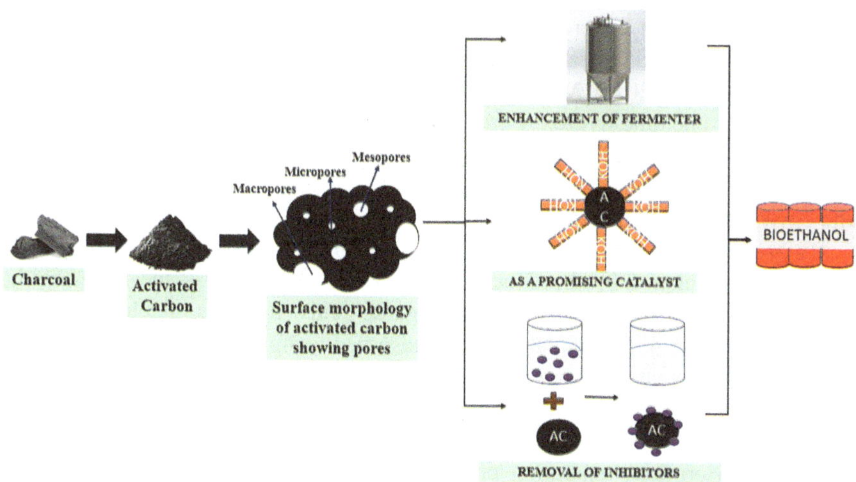

Fig. 7.5 Fates of activated carbon in bioethanol production

7.4.2 Purification of Hydroxylate

Recently, pervapouration membranes for possible bioethanol recovery were examined and addressed (Peng et al. 2021). Additionally, membrane systems for the detoxification of hydrolysates were investigated. Specifically, particular inhibitors were removed using nanofiltration, diafiltration, and pervapouration, preventing the loss of sugars. Due to the high porosity of the substance, which is also being investigated for the detoxification of lignocellulosic hydrolysates, activated carbon adsorption is frequently used for purification.

The activated charcoal particles have the ability to bind and hold contaminants through surface contacts when added to a hydroxylate solution, effectively eliminating them from the mixture. It preferentially absorbs impurities from the hydroxylate solution, including organic pollutants and certain metal ions. To achieve complete contact between the charcoal particles and the solution, the activated charcoal is combined with the hydroxylate solution and frequently stirred. The hydrolysates were kept at 4 °C and normally filtered or centrifuged to isolate the activated charcoal for purification. The production of ethanol and butanol from lignocellulosic biomass requires the removal of inhibitory substances produced during hydrothermal or thermochemical pretreatment of the biomass, such as furfural, hydroxymethylfurfural (HMF), coumaric acid, syringic acid, vanillin, vanillic acid, and cinnamaldehyde, which can severely inhibit enzymatic hydrolysis and subsequent microbial fermentation (Klasson et al. 2013; Lee et al. 2011; Liu et al. 2015a, 2015b).

7.4.3 Adsorption of By-products

A promising method for creating effective adsorbent at cheap cost with enriched functional groups and increased surface area is the manufacture of biochar from rice straw (Khan et al. 2019; Melia et al. 2019). Aliphatic double bonds, hydroxyl, carbonyl, and phenolic hydroxyl groups, as well as the benzene ring, are the main functional groups found in RSB. These functional groups have given the RSB qualities such a high capacity for adsorption, resistance to oxidation, hydrophobicity, and environmental stability (Zhang et al. 2019).

By-products and contaminants from the bioethanol production process may build up in the fermentation broth, reducing the process' effectiveness and lowering the ethanol's quality. By-products such as organic acids, furfurals, aldehydes, heavy metal ash, soluble pollutants, etc. can be adsorbed using activated charcoal. These by-products are created during the breakdown of lignocellulosic biomass or result from microbial fermentation metabolism. Because of the selective nature of the adsorption process, activated charcoal has an affinity for particular substances. Following adsorption, separation procedures are used to separate the mixture of activated charcoal and adsorbed by-products.

In general, lignocellulosic materials used to make monosaccharide-rich hydrolysates also produce a variety of other by-products, such as furfural and related chemicals, during the chemical hydrolysis process. Pentose breakdown produces furfural chemicals, which are extremely hazardous to species that ferment sugars. Aside from acetic acid and phenolic chemicals, other by-products can also prevent fermentation (Zhang et al. 2011). Optimizing elements like the dosage of the charcoal, the contact time, the temperature, and the pH is necessary for the successful use of activated charcoal adsorption. It can result in increased ethanol yields and, consequently, better product quality in the bioethanol manufacturing process.

7.4.4 Enhancement of Fermentation

Since activated charcoal has the capacity to eliminate inhibitory compounds and absorb by-products, it aids in promoting an environment that is more favourable to microorganisms. This encourages their development, metabolism, and ethanol production, which increases yields and boosts fermentation efficiency to a greater extent. At multiple stages during the fermentation process, such as during pretreatment, hydrolysis, or right into the fermentation tank, activated charcoal can be added. Fermentation inhibitory substances, such as aliphatic acids, aromatic compounds, and derivatives of furans, are produced during pretreatment and are mostly determined by two factors:

1. The chemical composition of the lignocellulosic feedstock
2. The degree of the pretreatment technique used

Fermentation inhibitors such as aliphatic acids like acetic acid and furan aldehydes (furfural and HMF) are frequently present in pretreatment lignocellulosic hydrolysates (Sarawan et al. 2019). In order of decreasing toxicity, the following substances are toxic to the bioethanol fermentation system (Mussatto and Roberto 2004):

$$CH_3COOH > \text{extractives} > \text{Aromatic compounds} > \text{furfural 5} \\ - \text{Hydroxymethylfurfural (HMF)}$$

In one such study, lignocellulosic biomass-derived biochar was successfully used to immobilize the fermentative microbes, *S. cerevisiae*, *Kluyveromyces marxianus*, and *Pichia kudriavzevii* KVMP10, and the research findings revealed that the immobilization of the yeast cells in repeated batch fermentations improved bioethanol production (Kyriakou et al. 2019). Remarkably a bit of research has previously been done on biochar made from lignocellulosic biomass. Many environmental pollutants, such as organic, inorganic, and gas pollutants, as well as heavy metals, have been successfully eliminated using biochar. According to a study, the biochar in their self-sufficient system had high adsorption efficiencies for the fermentation inhibitors vanillin, syringaldehyde, and 4-hydroxybenzaldehyde, respectively, of 98.6%, 98.15%, and 98.2%. Additionally, the biochar in their system had a lower rate of lowering sugar adsorption (5.6%).

7.4.5 *As a Promising Catalyst*

The excellent thermal conductivity, meso-/microporous structure, and chemical and thermal stability of ACs have made them the focus of research into their attributes. The primary advantage of AC compared to alternative materials lies in the easy availability of cost-effective source materials, facilitating the production of a diverse range of AC catalysts. A lot of functional groups are present on the surface of AC, which has a surface area of up to 1500 m^2/g.

The elevated catalytic efficacy of AC is attributed to its acidic surface and valuable surface characteristics, enabling its utilization as a catalyst support. The O-surface functional groups act as sites for securing metallic substances, highlighting AC's attributes as a material for supporting catalysts. Lignocellulosic biomass has been successfully hydrolysed and dehydrated using biochar-based catalysts (Xiong et al. 2018). With the help of a charcoal catalyst, cellulose and hemicellulose were hydrolysed, cleaving the glycosidic linkages in polysaccharides like glucan and xylan to produce reducing sugars like glucose and xylose. By using a sulphonated charcoal catalyst, these sugars can be further dehydrated to produce furfural and HMF. Biochar-based catalysts are less expensive, are easier to separate for reuse, and have less environmental problems than enzymes or acids and bases (Sun et al. 2020). It has also been widely used as an efficient catalyst in the manufacturing of biodiesel, as well as for gas separation, solvent recovery, and wastewater treatment (Tavares et al. 2022).

7.5 Conclusion

The fabrication and utilization of activated charcoal in bioethanol production present a promising avenue towards enhancing the efficiency, sustainability, and economic viability of the biofuel industry. Through a comprehensive exploration of the properties, preparation methods, and applications of activated charcoal, it becomes evident that this versatile adsorbent holds immense potential in addressing the challenges associated with ethanol purification. Activated charcoal's high surface area, porosity, and selective adsorption capabilities make it a valuable tool for removing impurities, colourants, and water from bioethanol-water mixtures. Its regenerability adds a crucial dimension of sustainability, aligning with the principles of circular economy and resource optimization. Moreover, the utilization of agricultural residues, such as coconut shells, sawdust, or bamboo, as precursors for activated charcoal fabrication contributes to waste reduction and the valorization of otherwise underutilized materials. The integration of activated charcoal into bioethanol production processes can lead to several positive outcomes. It can enhance the overall ethanol yield by improving the purity of the final product, minimizing energy-intensive distillation requirements, and reducing losses during purification. Furthermore, activated charcoal's potential to mitigate environmental impacts through the adsorption of pollutants and its regenerable nature contribute to a greener and more sustainable bioethanol industry. However, challenges persist, such as optimizing activation parameters, tailoring activated charcoal properties to specific feedstocks, and assessing the economic feasibility of large-scale production. Furthermore, the broader adoption of activated charcoal in the industry requires collaboration among researchers, policymakers, and stakeholders to ensure seamless integration into existing production processes.

In conclusion, the fabrication of activated charcoal for bioethanol production offers a transformative approach to address the purification challenges of bioethanol. Its unique characteristics, renewable precursor materials, and versatile applications underscore its potential to revolutionize the biofuel landscape. As the world seeks sustainable energy alternatives, activated charcoal stands as a testament to the power of innovation and resourcefulness in achieving a more environmentally conscious and economically viable future.

References

Abd AA, Naji SZ, Hashim AS, Othman MR (2020) Carbon dioxide removal through physical adsorption using carbonaceous and non-carbonaceous adsorbents: a review. J Environ Chem Eng 8(5):104142

Abd AA, Othman MR, Kim J (2021) A review on application of activated carbons for carbon dioxide capture: present performance, preparation, and surface modification for further improvement. Environ Sci Pollut Res 28(32):43329–43364

Abd-Rabboh HS, Fawy KF, Hamdy MS, Elbehairi SI, Shati AA, Alfaifi MY, Ibrahium HA, Alamri S, Awwad NS (2022) Valorization of rice husk and straw agriculture wastes of Eastern Saudi Arabia: production of bio-based silica, lignocellulose, and activated carbon. Materials 15(11):3746

Aro T, Fatehi P (2017) Production and application of lignosulfonates and sulfonated lignin. ChemSusChem 10(9):1861–1877

Byamba-Ochir N, Shim WG, Balathanigaimani MS, Moon H (2016) Highly porous activated carbons prepared from carbon rich Mongolian anthracite by direct NaOH activation. Appl Surf Sci 379:331–337

Cheang TY, Tang B, Xu AW, Chang GQ, Hu ZJ, He WL, Xing ZH, Xu JB, Wang M, Wang SM (2012) Promising plasmid DNA vector based on APTES-modified silica nanoparticles. Int J Nanomedicine 7:1061–1067

Chen WH, Lin BJ, Lin YY, Chu YS, Ubando AT, Show PL, Ong HC, Chang JS, Ho SH, Culaba AB, Pétrissans A (2021) Progress in biomas torrefaction: principles, applications and challenges. Prog Energy Combust Sci 82:100887

Do NH, Pham HH, Le TM, Lauwaert J, Diels L, Verberckmoes A, Do NH, Tran VT, Le PK (2020) The novel method to reduce the silica content in lignin recovered from black liquor originating from rice straw. Sci Rep 10(1):21263

Duarte LC, Carvalheiro F, Neves I, Gírio FM (2005) Effects of aliphatic acids, furfural, and phenolic compounds on debaryomyces hansenii CCMI 941. Appl Biochem Biotechnol 121: 413–425

El-Sayed MA, El-Samni TM (2006) Physical and chemical properties of rice straw ash and its effect on the cement paste produced from different cement types. J King Saud Univ Eng Sci 19(1): 21–29

Estévez MC, O'Donoghue MB, Chen X, Tan W (2009) Highly fluorescent dye-doped silica nanoparticles increase flow cytometry sensitivity for cancer cell monitoring. Nano Res 2(6): 448–461

Farooqui A, Tripathi G, Moheet K, Dubey P, Ahmad S, Husain A, Shamim A, Mahfooz S (2021) Algal biomass: potential renewable feedstock for bioenergy production. In: Srivastava M, Srivastava N, Singh R (eds) Bioenergy research: integrative solution for existing roadblock. Springer, Singapore, pp 85–113

Galliker P, Hommes G, Schlosser D, Corvini PFX, Shahgaldian P (2010) Laccase-modified silica nanoparticles efficiently catalyze the transformation of phenolic compounds. J Colloid Interface Sci 349(1):98–105

Gao P, Liu ZH, Xue G, Han B, Zhou MH (2011) Preparation and characterization of activated carbon produced from rice straw by (NH_4) $2HPO_4$ activation. Bioresour Technol 102(3): 3645–3648

Gupta PA, Sharma V, Gade MK (2015) Rice husk power systems: exploring alternate source of energy. In: Promoting socio-economic development through business integration. IGI Global, Hershey, PA, pp 112–123

Gupta R, Gautam S, Shukla R, Kuhad RC (2017) Study of charcoal detoxification of acid hydrolysate from corncob and its fermentation to xylitol. J Environ Chem Eng 5(5):4573–4582

Hans M, Kumar S, Chandel AK, Polikarpov I (2019) A review on bioprocessing of paddy straw to ethanol using simultaneous saccharification and fermentation. Process Biochem 85:125–134

Heidarinejad Z, Dehghani MH, Heidari M, Javedan G, Ali I, Sillanpää M (2020) Methods for preparation and activation of activated carbon: a review. Environ Chem Lett 18:393–415

Khan MN, Lan Z, Sial TA, Zhao Y, Haseeb A, Jianguo Z, Zhang A, Hill RL (2019) Straw and biochar effects on soil properties and tomato seedling growth under different moisture levels. Arch Agron Soil Sci 65(12):1704–1719

Klasson KT, Dien BS, Hector RE (2013) Simultaneous detoxification, saccharification, and ethanol fermentation of weak-acid hydrolyzates. Ind Crop Prod 49:292–298

Kumari A, Kapoor KK, Kundu BS, Kumari Mehta R (2008) Identification of organic acids produced during rice straw decomposition and their role in rock phosphate solubilization. Plant Soil Environ 54(2):72

Kwon S, Singh RK, Perez RA, Abou Neel EA, Kim HW, Chrzanowski W (2013) Silica-based mesoporous nanoparticles for controlled drug delivery. J Tissue Eng 4:2041731413503357

Kyriakou M, Chatziiona VK, Costa CN, Kallis M, Koutsokeras L, Constantinides G, Koutinas M (2019) Biowaste-based biochar: a new strategy for fermentative bioethanol overproduction via whole-cell immobilization. Appl Energy 242:480–491

Lee JM, Venditti RA, Jameel H, Kenealy WR (2011) Detoxification of woody hydrolyzates with activated carbon for bioconversion to ethanol by the thermophilic anaerobic bacterium thermoanaerobacterium saccharolyticum. Biomass Bioenergy 35(1):626–636

Liu Y, Tourbin M, Lachaize S, Guiraud P (2013) Silica nanoparticles separation from water: aggregation by cetyltrimethylammonium bromide (CTAB). Chemosphere 92(6):681–687

Liu K, Atiyeh HK, Pardo-Planas O, Ezeji TC, Ujor V, Overton JC, Berning K, Wilkins MR, Tanner RS (2015a) Butanol production from hydrothermolysis-pretreated switchgrass: quantification of inhibitors and detoxification of hydrolyzate. Bioresour Technol 189:292–301

Liu K, Atiyeh HK, Pardo-Planas O, Ramachandriya KD, Wilkins MR, Ezeji TC, Ujor V, Tanner RS (2015b) Process development for biological production of butanol from Eastern redcedar. Bioresour Technol 176:88–97

Logeswaran J, Shamsuddin AH, Silitonga AS, Mahlia TMI (2020) Prospect of using rice straw for power generation: a review. Environ Sci Pollut Res 27:25956–25969
Lu P, Hsieh YL (2012) Highly pure amorphous silica nano-disks from rice straw. Powder Technol 225:149–155
Lu J, Liong M, Zink JI, Tamanoi F (2007) Mesoporous silica nanoparticles as a delivery system for hydrophobic anticancer drugs. Small 3(8):1341–1346
Mazlan MAF, Uemura Y, Yusup S, Elhassan F, Uddin A, Hiwada A, Demiya M (2016) Activated carbon from rubber wood sawdust by carbon dioxide activation. Proc Eng 148:530–537
Melia PM, Busquets R, Hooda PS, Cundy AB, Sohi SP (2019) Driving forces and barriers in the removal of phosphorus from water using crop residue, wood and sewage sludge derived biochars. Sci Total Environ 675:623–631
Mussatto SI, Roberto IC (2004) Alternatives for detoxification of diluted-acid lignocellulosic hydrolyzates for use in fermentative processes: a review. Bioresour Technol 93(1):1–10
Naji SZ, Tye CT (2022) A review of the synthesis of activated carbon for biodiesel production: precursor, preparation, and modification. Energy Conv Manage X 13:100152
Nandiyanto ABD, Ogi T, Iskandar F, Okuyama K (2011) Highly ordered porous monolayer generation by dual-speed spin-coating with colloidal templates. Chem Eng J 167(1):409–415
Nandiyanto ABD, Suhendi A, Ogi T, Umemoto R, Okuyama K (2014) Size-and charge-controllable polystyrene spheres for templates in the preparation of porous silica particles with tunable internal hole configurations. Chem Eng J 256:421–430
Nandiyanto ABD, Rahman T, Fadhlulloh MA, Abdullah AG, Hamidah I, Mulyanti B (2016) Synthesis of silica particles from rice straw waste using a simple extraction method. IOP Conf Ser Mater Sci Eng 128(1):012040
Narowska BE, Kułażyński M, Łukaszewicz M (2020) Application of activated carbon to obtain biodiesel from vegetable oils. Catalysts 10(9):1049
Nasr AI, Taha MG, Yosef HY, El-Shaer MA (2017) The utilization of sodium lignosulphonate extracted from Egyptian rice straw in leather tanning process. Majalah Kulit Karet dan Plastik 33(2):57–64
Nolan A, Mc Donnell K, Devlin GJ, Carroll JP, Finnan J (2010) Economic analysis of manufacturing costs of pellet production in the Republic of Ireland using non-woody biomass. Open Renew Energy J 3(1):1–11
Norrahim MNF, Kasim NAM, Knight VF, Ujang FA, Janudin N, Razak MAIA, Shah NAA, Noor SAM, Jamal SH, Ong KK, Yunus WMZW (2021) Nanocellulose: the next super versatile material for the military. Mater Adv 2(5):1485–1506
Pallarés J, González-Cencerrado A, Arauzo I (2018) Production and characterization of activated carbon from barley straw by physical activation with carbon dioxide and steam. Biomass Bioenergy 115:64–73
Peng P, Lan Y, Liang L, Jia K (2021) Membranes for bioethanol production by pervaporation. Biotechnol Biofuels 14(1):1–33
Phiri Z, Everson RC, Neomagus HW, Wood BJ (2017) The effect of acid demineralising bituminous coals and de-ashing the respective chars on nitrogen functional forms. J Anal Appl Pyrolysis 125:127–135
Pulicharla R, Lonappan L, Brar SK, Verma M (2016) Production of renewable C_5 platform chemicals and potential applications. In: Brar SK, Sarma SJ, Pakshirajan K (eds) Platform chemical biorefinery. Elsevier, Amsterdam, pp 201–216
Purohit P, Chaturvedi V (2018) Biomass pellets for power generation in India: a techno-economic evaluation. Environ Sci Pollut Res 25:29614–29632
Pushpa J, Sinha A (2011) Application of rice straw as raw material for the production of handmade paper. IPPTA 23(2):145–148
Rao KS, Mahender C, Ramulu DV (2022) Production of activated carbon from rice husk and rice straw for adsorption studies. Int J Innov Sci Res Technol 7(9):156–158
Roy GM (2023) Activated carbon applications in the food and pharmaceutical industries. Routledge, New York

Sadeek SA, Mohammed EA, Shaban M, Abou Kana MT, Negm NA (2020) Synthesis, characterization and catalytic performances of activated carbon-doped transition metals during biofuel production from waste cooking oils. J Mol Liq 306:112749

Sahri Y, Tamalouzt S, Belaid SL (2018) Direct torque control of DFIG driven by wind turbine system connected to the grid. In 2018 international conference on wind energy and applications in Algeria (ICWEAA), IEEE, November, pp 1–6

Sakhiya AK, Vijay VK, Kaushal P (2022) Efficacy of rice straw derived biochar for removal of Pb^{+2} and Zn^{+2} from aqueous: adsorption, thermodynamic and cost analysis. Bioresour Technol Rep 17:100920

Sarawan C, Suinyuy TN, Sewsynker-Sukai Y, Kana EG (2019) Optimized activated charcoal detoxification of acid-pretreated lignocellulosic substrate and assessment for bioethanol production. Bioresour Technol 286:121403

Sharath D, Ezana J, Shamil Z (2017) Production of activated carbon from solid waste rice peel (husk) using chemical activation. J Ind Pollut Control 33(2):1132–1139

Singh L, Brar BS (2021) A review on rice straw management strategies. Nat Environ Pollut Technol 20(4):1485–1493

Singh R, Kumar S (2019) A review on biomethane potential of paddy straw and diverse prospects to enhance its biodigestibility. J Clean Prod 217:295–307

Singh RB, Saha RC, Singh M, Chandra D, Shukla SG, Walli TK, Pradhan PK, Kessels HPP (1995) Rice straw, its production and utilization in India. In: Handbook for straw feeding systems in livestock production. ICAR, New Delhi, pp 325–339

Singh LP, Agarwal SK, Bhattacharyya SK, Sharma U, Ahalawat S (2011) Preparation of silica nanoparticles and its beneficial role in cementitious materials. Nanomater Nanotechnol 1:9

Singh R, Srivastava M, Shukla A (2016) Environmental sustainability of bioethanol production from rice straw in India: a review. Renew Sust Energ Rev 54:202–216

Singh P, Dubey P, Younis K, Yousuf O (2022) A review on the valorization of coconut shell waste. Biomass Conv Bioref. https://doi.org/10.1007/s13399-022-03001-2

Soleimani M, Tabil L, Niu C (2015) Adsorptive isotherms and removal of microbial inhibitors in a bio-based hydrolysate for xylitol production. Chem Eng Commun 202(6):787–798

Sun X, Liu C, Omer AM, Lu W, Zhang S, Jiang X, Wu H, Yu D, Ouyang XK (2019) pH-sensitive ZnO/carboxymethyl cellulose/chitosan bio-nanocomposite beads for colon-specific release of 5-fluorouracil. Int J Biol Macromol 128:468–479

Sun X, Atiyeh HK, Li M, Chen Y (2020) Biochar facilitated bioprocessing and biorefinery for productions of biofuel and chemicals: a review. Bioresour Technol 295:122252

Tadda MA, Ahsan A, Shitu A, ElSergany M, Arunkumar T, Jose B, Razzaque MA, Daud NN (2016) A review on activated carbon: process, application and prospects. J Adv Civil Eng Pract Res 2(1):7–13

Tamalouzt S, Rekioua T, Abdessemed R (2014). Direct torque and reactive power control of grid connected doubly fed induction generator for the wind energy conversion. In: 2014 international conference on electrical sciences and technologies in Maghreb (CISTEM), IEEE, pp 1–7

Tavares AP, Gonçalves MJ, Brás T, Pesce GR, Xavier AM, Fernandes MC (2022) Cardoon hydrolysate detoxification by activated carbon or membranes system for bioethanol production. Energies 15(6):1993

Tripathi G, Jamal A, Jamal T, Faiyaz M, Farooqui A (2022) Phyco-nanotechnology: an emerging nanomaterial synthesis method and its applicability in biofuel production. In: Srivastava M, Malik MA, Mishra PK (eds) Green nano solution for bioenergy production enhancement. Springer, Singapore, pp 169–200

Xiong X, Iris KM, Chen SS, Tsang DC, Cao L, Song H, Kwon EE, Ok YS, Zhang S, Poon CS (2018) Sulfonated biochar as acid catalyst for sugar hydrolysis and dehydration. Catal Today 314:52–61

Yahya MA, Al-Qodah Z, Ngah CZ (2015) Agricultural bio-waste materials as potential sustainable precursors used for activated carbon production: a review. Renew Sust Energ Rev 46:218–235

Yang K, Peng J, Srinivasakannan C, Zhang L, Xia H, Duan X (2010) Preparation of high surface area activated carbon from coconut shells using microwave heating. Bioresour Technol 101(15): 6163–6169
Yanti Y, Surahmanto S, Purnomoadi A, Kawamoto Y (2012) Organic acids production of rice straw fermented with several types of microorganism at different temperatures. J Indones Trop Anim Agric 37(3):189–194
Yorgun S, Yıldız D, Şimşek YE (2016) Activated carbon from paulownia wood: yields of chemical activation stages. Energy Sources A Recov Util Environ Effects 38(14):2035–2042
Yu W, Dong L, Lei W, Shi J (2020) Rice straw powder/polylactic acid biocomposites for three-dimensional printing. Adv Compos Lett 29:2633366X20967360
Zhang K, Agrawal M, Harper J, Chen R, Koros WJ (2011) Removal of the fermentation inhibitor, furfural, using activated carbon in cellulosic-ethanol production. Ind Eng Chem Res 50(24): 14055–14060
Zhang Y, Yue X, Xu W, Zhang H, Li F (2019) Amino modification of rice straw-derived biochar for enhancing its cadmium (II) ions adsorption from water. J Hazard Mater 379:120783
Zhong ZY, Yang Q, Li XM, Luo K, Liu Y, Zeng GM (2012) Preparation of peanut hull-based activated carbon by microwave-induced phosphoric acid activation and its application in Remazol Brilliant Blue R adsorption. Ind Crop Prod 37(1):178–185

Chapter 8
Bioethanol Production from Paddy Straw Lignocellulosic Waste

Bushra Iqbal, Misbah Ghazanfar, Hafiz Abdullah Shakir, Shaukat Ali, Muhammad Khan, Alia Gul, Marcelo Franco, and Muhammad Irfan

Abstract A viable method for using trash as a source of waste management and sustainable energy is the synthesis of bioethanol from lignocellulosic waste such as paddy straw. Complex carbohydrates found in paddy straw, a common agricultural waste product, may be processed into ethanol in a number of steps. An overview of the generation of bioethanol from lignocellulosic waste using paddy straw is given in this chapter. Pretreatment is the initial phase in the procedure, when the lignocellulosic structure of the paddy straw is broken down using methods like steam explosion, acid hydrolysis, or alkaline treatment. The carbohydrate accessibility for later enzymatic hydrolysis is improved by this pretreatment. Following enzymatic hydrolysis, enzymes like cellulases and hemicellulases convert the components of cellulose and hemicellulose into fermentable sugars. The resulting sugar solution, which comprises glucose, xylose, and other monomeric sugars, is then fermented. *Saccharomyces cerevisiae*, a common yeast strain, has the capacity to transform glucose into ethanol. Generally speaking, the creation of bioethanol using lignocellulosic waste from paddy straw offers a potential solution for the management of agricultural leftover and the production of renewable fuels. Paddy straw has a lot of potential as a feedstock for production of bioethanol; thus ongoing research is working to make the processes more effective, save costs, and create new technologies.

Keywords Paddy straw · Pretreatment · Ethanol · Biofuel · Biomass · RS

B. Iqbal · M. Ghazanfar · M. Irfan (✉)
Faculty of Science, Department of Biotechnology, University of Sargodha, Sargodha, Pakistan
e-mail: irfan.ashraf@uos.edu.pk

H. A. Shakir · S. Ali · M. Khan
Institute of Zoology, University of the Punjab, Lahore, Pakistan

A. Gul
Department of Botany, Hazara University, Mansehra, Pakistan

M. Franco
Department of Exact Science and Technology, State University of Santa Cruz, Ilhus, Brazil

N. Srivastava et al. (eds.), *Paddy Straw Waste for Biorefinery Applications*, Clean Energy Production Technologies, https://doi.org/10.1007/978-981-99-8224-0_8

8.1 Introduction

Rice straw (RS) is one of the lignocellulosic waste products that is widely dispersed globally. Although it has a lower net yield than maize and wheat, rice is nonetheless an important grain crop worldwide. In 2007, the globe produced about 650 million tons of rice annually, according to FAO figures. According to Maiorella (1985), 1.1–1.5 kg of straw are generated for every kilogram of harvested grain. Across 650–975 million tons of RS are generated annually across the world; the bulk of this material is discarded and utilized as cow fodder. Due to its high mineral content, vulnerability to rice stem diseases, gradual topsoil deterioration, and sparse bulk density, there are few options for disposing of RS. Despite the fact that it worsens air pollution and endangers public health, field burning is now the main method used to remove rice straw (Mussatto and Roberto 2004). Waste from the agriculture sector is increasingly being used for energy-related purposes as the danger posed by climate change to growth becomes more widely acknowledged. If true, using RS to meet our energy demands in the future could prove to be a good decision. This chapter provides a comprehensive review of the method for producing bioethanol that is compatible with rice straw.

8.1.1 RS: Availability and Composition

In India, the Kharif harvest season results in a significant volume of rice straw waste each year (Fig. 8.1). Planning how to use RS, either alone or in combination with other biomasses, requires a thorough understanding of its chemical composition. However, depending on the kind, season, climatic circumstances, and geographic regions where rice is grown, the general makeup of this biomass may occasionally change unfavorably. In addition to cellulose and hemicellulose, RS also includes considerable amounts of lignin, silica, and other minerals. Chemically, the bulk of RS is composed of 32% to 47% cellulose, 19% to 27% hemicellulose, and 5% to 24% lignin (Binod et al. 2010; Takkellapati et al. 2018). In order to provide the cell wall of plants, a particularly strong structure of lignin is not found in plants as a single molecule; rather, it forms a complex with phenolics, carbohydrates, and other compounds. In terms of the carbohydrates present in its cell walls, the rice plant nevertheless has great heterogeneity: −27% uronic acids, 63% lignin, 4% trans-ferulic acid, 4% trans-p-coumaric acid, and −2% acetyl content (1%) (Lewis and Yamamoto 1990).

RS has both esterified and etherified ferulic acid and p-coumaric acid in larger amounts. Despite having lignified, hemicellulose, and cellulosic cell walls in addition to low content of nitrogen, RS includes a significant quantity of ash approximately 10% that has a high amount of silica (containing an estimated 75% SiO_2 and less than 15% alkali content) (Saritha et al. 2013). Rice was among the earlier grain crops that human cultivated. Nevertheless, the origins of rice, particularly the *Oryza*

Fig. 8.1 Collected RS

sativa cultivation, are mainly linked to Asia (Callaway 2014). It is possible to contest the origins of rice and the theory that it was domesticated in China by people of the Yangtze River valley between 8000 and 10,000 BCE by recounting the words of Dorian Fuller, a renowned archaeobotanist, "Asian civilization was established on rice-on Oryza sativa, to be exact." Some more modern kinds, however, might be attributed to various periods and locations around the globe like *O. indica*, *O. glaberrima*, and *O. japonica*. *O. sativa*, which originated in Asia, and *O. glaberrima*, which originated in Africa, are the two widely produced and starkly different types of rice. Although their domestication is highly debated, they have widely cultivated varieties (Sweeney and McCouch 2007). Almost everywhere in the world, rice is predominantly a significant edible, demonstrating its fundamental significance in determining the past, present, and future of humanity. The high productivity of rice and government incentives for farmers are projected to lead to future increases in rice acreage as well. There will also be more RS available, which can be processed using the right technologies to create biofuel, fertilizer, and other value-added goods.

8.1.2 Ethanol Fuel Production Potential of RS

The utilization of biomass-derived ethanol as a gasoline substitute is growing in marketability. Unfortunately, the first generation of biofuels, bioethanol, which is made from grains, has an unfavorable head-on conflict with the provision of food. A change to a prevalent non-edible vegetative matter ought to help ease the strain on the food crops. These plant components contain significant amounts of fermentable sugar-producing hemicellulose and cellulose. Bacteria may use these carbohydrates for fermentation to produce ethanol. RS is a suitable feedstock for ethanol production due to a number of factors. These include large amounts of cellulose and hemicellulose, which are quickly transformed into fermentable sugars. Hemicellulose (19–27%), cellulose (3–47%), and lignin (5%) are straw's primary chemical constituents (Garrote et al. 2002; Maiorella 1983; Saha 2003; Zamora and Crispin 1995). While xylose takes from 14.8% to 20.2% of the total sugar content in hemicellulose, pentoses predominate in this material (Maiorella 1983; Roberto et al. 2003). Table 8.1 displays the theoretic production of bioethanol and carbohydrates from RS. The chemical makeup of the substrate has a considerable influence on how well bioenergy is produced. The chemical properties of rice husk, wheat straw, and RS are provided in the second table to highlight the unique variances in feedstock. When compared to wheat straw, RS has a higher percentage of ash (10–17% vs. 3%), and it has a higher silica concentration (75% vs. 55%, respectively) (Zevenhoven 2000). RS has a somewhat lower overall alkali concentration than wheat straw (K_2O and Na_2O typically make around 15% of the total ash), which frequently contains more than 25% alkali (Baxter et al. 1996).

The fractions of hemicellulose are presumed to be xylose polymers (Table 8.2).

Throughout the year and in different areas, straw quality varies greatly. Alkaline and alkali chemicals are released from a straw in the field when it is subjected to precipitation, raising the feedstock's quality. Because of this, the recommended way to use straw for making bioethanol relies on both its quality and availability.

8.1.3 RS's Potential as a Green Biorefinery

As indicated in the introduction, traditionally, waste biomass produced during rice processing was inflamed or utilized for fertilizing, flooring, and roofing, food for animals, and some handiwork (Matsumura et al. 2005). Modernization, however,

Table 8.1 Theoretical RS ethanol yield and carbohydrate breakdown

Hemicellulose	19.7%
Cellulose	38.6%
Theoretical ethanol yield (L/kg dry)	0.42
Theoretical ethanol yield (gal/MT dry)	110

Source: This information was taken from Zhu et al. (2005a, 2005b)'s Biomass Feedstock Composition and Property Database

Table 8.2 The approximate composition of a few key ash components of wheat straw, rice husk, and straw (Jenkins et al. 1998)

	Wheat straw	RS	Rice husk
Proximate analysis (% dry fuel)			
Volatile matter	75.27	65.47	63.52
Ash	7.02	18.67	20.26
Fixed carbon	17.71	15.86	16.22
Elemental composition of ash (%)			
K_2O	25.60	12.30	3.71
Na_2O	1.71	0.96	0.21
CaO	6.14	3.01	3.21
MgO	1.06	1.75	<0.01
SiO_2	55.32	74.67	91.42

has altered all of the concepts, and in recent years, reports of these leftovers as a possible feedstock for biorefineries have been made. A biorefinery is a method for effectively using biomass to create products with a marketable added value (Abo-State et al. 2014). The concepts for biorefineries follow a sequential order. The determination of the resulting compounds' molecular makeup and the creation of pathways for reactions for the maximal production of the required stock come after the initial consideration of social demands and potential applications of the products. The choice of appropriate feedstock for the production of the greatest possible number of compounds is the most crucial phase in this process. Most reported fine chemicals, biofuels, biopolymers, and organic acids come from biomass feedstocks ample in carbohydrates (Abo-State et al. 2014; Matsumura et al. 2005; Song et al. 2012). Biomass pyrolysis is the process that is primarily utilized to produce all types of fuels from these leftovers, including a liquid (bio-oil), a combination of light gases (syngas), and a solid (bio-char) (Saritha et al. 2013). As indicated in Fig. 8.2, RS can also be used to make various valuable goods, such as alcohols (via microbial fermentation), enzymes, proteins, bioinsecticides, biocomposites, hydrogen, coal, lactic acid, bio foam, biogas, furfurals, wood additives, and compost. The concentration and generation of these several final products, however, are significantly impacted by the normal technique and its factors (Abo-State et al. 2014).

According to reports (Abo-State et al. 2014; Saritha et al. 2013), RS was used in biorefineries to create sustainable fuels, notably supplemental biofuels. Based on the compositional section's description that RS is mostly made of hemicellulose, lignin, and cellulose, different types of biofuels can be created from this biomass. According to a paper by Song et al. (2012), RS may be used to create biogas by means of digestion (in absence of oxygen) process. The use of RS as a substrate for bioenergy synthesis has increased in recent decades, and numerous researches have been done since then to substantiate this assertion (Harun et al. 2022; Matsumura et al. 2005; Song et al. 2012). The efficiency of RS as biomass for bioenergy generation was examined by Dash et al. using the fungus *Mucor circinelloides*; Takano and Hoshino (2018) employed RS as a raw material in another study to produce ethanol. Rice waste's potential as a preferred feedstock in biorefineries was

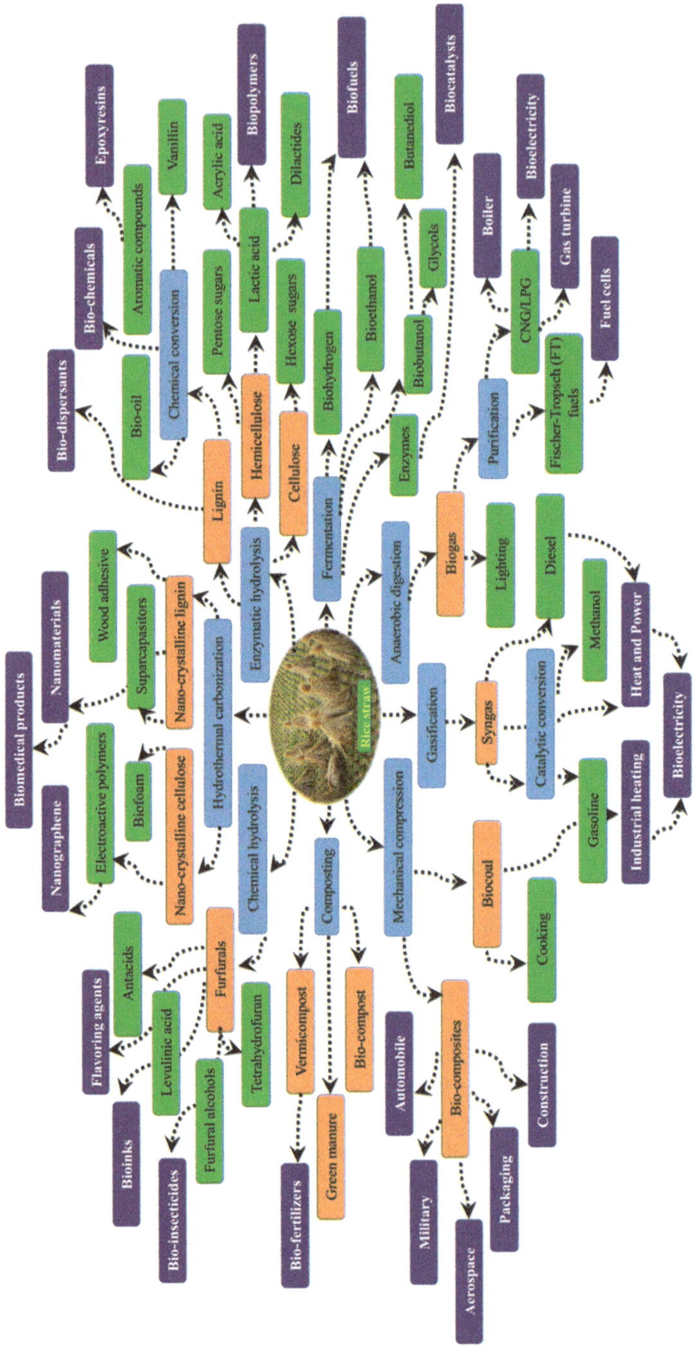

Fig. 8.2 Potential value-added products and their applications from biorefinery of RS (Taken from Rathour et al 2023)

assessed by Idris and Hashim (2021). Similar reports for composite creation from RS can be found (Idris and Hashim 2021; Takano and Hoshino 2018). Harun et al. (2022) emphasized the usefulness of RS in biorefineries in a recent study.

8.1.4 *Sustainable Extraction of Straw*

Because straw contains substantial nutrients, removing it from the field causes an imbalance in the humus and depletes the solid nutrients (Singh et al. 2016). According to Sheehan et al. (2003), straw removal from about 60% of the field does not affect soil quality or productivity. Koga and Tajima (2011) investigated if adding straw to the soil increased the amount of greenhouse gas emissions. Removing the straw from the field lowers the emissions of GHG.

8.2 Ethanol as Bioenergy Resources

The bioenergy conceptualization is an eco-friendly high-technology that integrates social, environmental, and economic concerns.

8.2.1 *Bioethanol from RS*

The most common method for reducing the straw's intricate carbohydrate structure to its basic form is enzymatic hydrolysis. However, a variety of elements such as moisture content, cellulose crystallinity, degree of polymerization (DP), lignin concentration, and accessible surface area limit the process (Laureano-Perez et al. 2005). So, pretreatment is crucial in the ethanol synthesis from straw process before fermentation and enzymatic hydrolysis.

8.3 RS Pretreatment

Pretreatment techniques may be broadly classified as chemical (diluted acid, alkali, organic solvents, and oxidizing agents,), physical (milling or grinding), physico-chemical (autohydrolysis, wet oxidation, hydro thermolysis, and steam pretreatment), biological, or a mixture of these (Fig. 8.3). Table 8.3 lists the specifics of the conditions for pretreatment and how they affect the biomass.

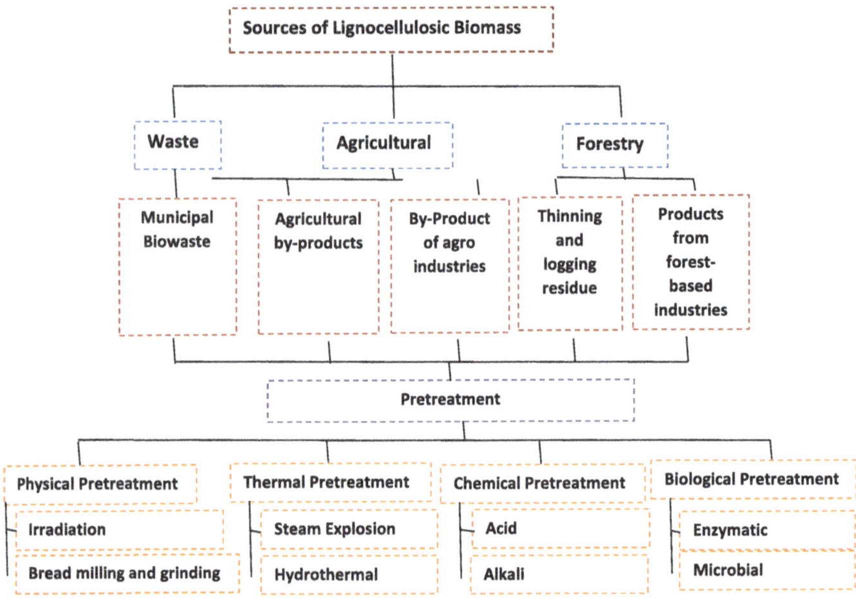

Fig. 8.3 Modes of pretreatment and different sources of lignocellulosic feedstock for bioethanol production

8.3.1 Physical Pretreatment

Physical pretreatment involves milling, grinding, steaming, heating, irradiation, and pressure that reduces the cellulose DP and crystallinity while expanding the pore size and available surface area. According to Zhu et al. (2005a, 2005b), microwave irradiation may alter the cellulose ultrastructure of cellulose, cause the breakdown of hemicelluloses and lignin in lignocellulosic substances to break down, and make them extra prone to enzymatic degradation (Abedinifar et al. 2009). According to Hideno et al. (2009), wet disc milling performs better than ball milling in terms of energy savings and glucose restoration.

8.3.2 Milling

By milling, the straw's particle size is reduced, lowering the degree of polymerization and increasing the amount of specific surface that is available (Chang and Holtzapple 2000; Palmowski and Muller 1999). Elements including operational expenses, capital expenses, scale-up potential, and equipment devaluation costs are crucial for mechanical pretreatment. Based on the biomass type, length of milling, the increase in shearing, and specific surface area are significant aspects

Table 8.3 Bioethanol production pretreatment conditions for the RS

Reagents	Conditions followed for the pretreatment	Advantages of pretreatment	Xylan recovery (%)	Glucan recovery (%)	References
Pleurotus ostreatus	Using *Pl. ostreatus*, RS with a 60% water content was pretreated for 60 days	Lignin removal up to 41%	52	83	Taniguchi et al. (2005)
Acid	RS and sulfuric acid in a 1:10 mixture, with a reaction time of 1–5 min at 160–180 °C and a residence period of 1 min	More than 90% xylan solubilized	4.7	57	Hsu et al. (2010)
Alkali and microwave	Following the application of 1% NaOH solution at a 1:8 solid to liquid ratio for periods of 15 to 2 h, the microwave was operated at a frequency of 2450 MHz and a power setting of 700, 500, and 300 W, respectively	Increased lignin and hemicellulose removal, as well as better hydrolytic enzyme accessibility	10.2	69.2	Zhu et al. (2005a, 2005b)
Phanerochaete chrysosporium	On RS, *Ph. chrysosporium* was grown for 15 days on response surface method-optimized medium	RS spontaneously loses lignin thanks to the ligninolytic enzyme that white rot fungus generate	10.9	37.7	Bak et al. (2009)
Dilute acid hydrolysis	After 20 h of soaking in 0.5% sulfuric acid at a solid-liquid ratio of 1:6.66, RS was added to the reactor, heated with stream heat for 1.5 min to reach pressure of	Removal of 99% of the hemicelluloses is necessary for enzymatic hydrolysis	1	55	Abedinifar et al. (2009)

(continued)

Table 8.3 (continued)

Reagents	Conditions followed for the pretreatment	Advantages of pretreatment	Xylan recovery (%)	Glucan recovery (%)	References
	15 bar, and then maintained at this pressure for 10 min to allow for hydrolysis to occur				
Alkali (calcium capturing by carbonation)	For 1 h under 120 °C, combine different amounts of rice and line with water while maintaining a solid to liquid ratio of 1:20	There is no need for a washing step or a solid-liquid separation phase for the solid part	81.5	95.2	Park et al. (2010)
Organosolv	75% (v/v) of RS was mixed with aqueous ethanol at a solid-to-liquid ratio of 1:8, with 1% (w/w) sulfuric acid serving as the catalyst. Treatments were conducted for 60 min at 180 °C in a high-pressure stainless-steel tank	Pretreatment with organosolv can be used to produce solvent from RS effectively	22.4	62.3	Amiri et al. (2014)
Ionic liquids	At a solid-liquid ratio of 1:20, ionic liquid was combined with RS before being baked for 24 h at 130 °C	The recovery rate for glucan and xylan is fairly high	76	79	Nguyen et al. (2010)
Alkali	At a solid-liquid ratio of 1:10, RS was combined with 10% (v/v) ammonia solution and heated to 100 °C for 6 h	Recovery rates for glucan and xylan are relatively high	52	84	Nguyen et al. (2010)
Reagents	**Pretreatment conditions**	**Pretreatment advantages**	**Recovery of xylan (%)**	**Recovery of glucan (%)**	**References**

(continued)

Table 8.3 (continued)

Reagents	Conditions followed for the pretreatment	Advantages of pretreatment	Xylan recovery (%)	Glucan recovery (%)	References
Acid	10% (w/v) substrate concentration of RS subjected to 2% (w/v) sulfuric acid hydrolysis under pressure for 30 min at 132 °C	Renders RS less crystallin and xylan-free, improving its suitability for saccharification	10.8	50.7	Oberoi et al. (2010)
Alkali	10% (w/v) RSt was autoclave-sterilized at 121 °C for 30 min after being treated with NaOH concentrations of 1–5% (w/v) and incubated at 40 °C for an hour	Lignin decreased by 47% and glucan increased by 50%	11.84	57.78	Oberoi et al. (2012)
Microwave alkali	8% (w/v) RS was microwaved for 30 min at 700 W after being prepared with 1% NaOH aqueous solution for 30 min under boiling conditions	Lignin decreased by 76% because of basic treatment and changes in the RS crystallinity	10.2	69.2	Zhu et al. (2005a, 2005b)
Beam radiation	Employing accelerated electrons from a linear electron accelerator to ionize RS. The straw spent the entire night in mineral water before being exposed to radiation	Improve the enzymatic digestibility	—	—	Bak (2014)
Microwave organic acid	A certain amount of acetic acid was combined with RS and microwaved for 2–5 min at 100–700 W	Lignin removal ratio was 51.54%	—	—	Gong et al. (2010)

that, in addition to increasing the hydrolysis rate by a 5–25% increase in the total lignocellulose production, also reduce the time for technical digestion by 23–59% (Delgenés et al. 2002).

Wet disc milling is an applied and inexpensive preparation for the saccharification of RS. It has been established that wet disc milling requires less energy than other comparable procedures (Hideno et al. 2009). Comparing ball milling to pretreatment by a steam explosion in wheat straw, ball milling is thought to be a more efficient pretreatment with a higher yield of glucose and carbohydrates.

8.3.3 *High-Energy Irradiation*

Several reports demonstrated that this procedure enhances accessibility to the crystalline portions of cellulose by increasing lignocellulose's crystallinity and lowering the polymerization degree (Bouchard et al. 2006) (Stepanik et al. 1998). As noted, pretreatment with electron beam treatment enhances the hydrolysis of RS by enzymes (Bak et al. 2009). To physically prepare milled, dry RS, Jin et al. (2009) used a linear electron accelerator to produce accelerated electrons in an electron beam irradiation. Using enzymes, RS that had been exposed to a beam of electrons and unprocessed RS generated 22.6% glucose yields and 52.1%, respectively, after being enzymatically hydrolyzed for 132 h. Water-soaked RS was exposed to irradiation by electron beam, according to a report by Bak (2014). When RS that had been bathed in aqueous ammonia was exposed to proton beam irradiation, glucose conversion was increased by up to 90%.

8.3.4 *Microwave Pretreatment*

In addition to altering the cellulose ultrastructure and hemicelluloses and lignin, microwave radiation may also make RS more susceptible to enzymatic degradation (Kitchaiya et al. 2003; Xiong et al. 2000). Its processing with alkali and microwave together has an effect on loss of weight (41.5–44.6%), due to hemicellulose and lignin elimination. Compared to an alkali-only hydrolysate, one produced using a microwave and an alkali had a greater rate of hydrolysis and a higher concentration of glucose (Zhu et al. 2005a, 2005b).

Likewise, Zhu et al. (2006) observed that using a microwave to treat wheat straw that had already been alkali pretreated increased the yield and ethanol content. The effect of alkali pretreatment by microwave on wheat straw was investigated for the ability to release saccharides and lignin. The solid wheat straw could have lignin (>90%) and hemicellulose (80%) removed from it without significant saccharide degrading or significant cellulose solubilization (Janker-Obermeier et al. 2012; Singh and Bishnoi 2012).

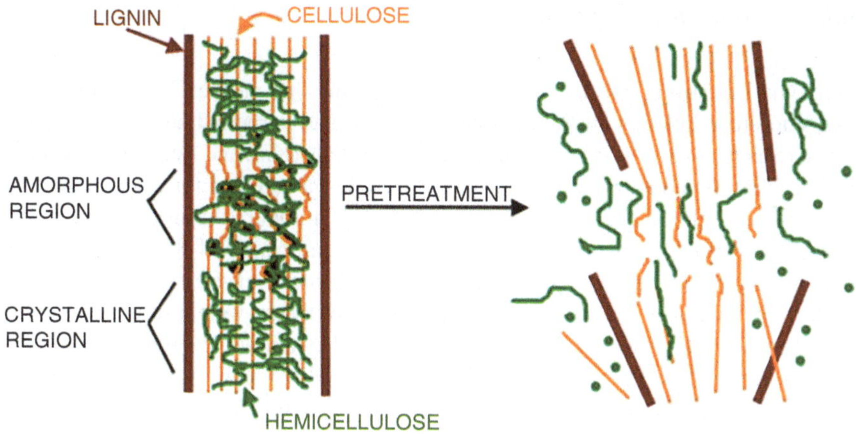

Fig. 8.4 Effect of pretreatment

8.3.5 Pretreatments Using Chemicals

Acids, alkali, peroxides ionic liquid (IL), and organic acid pretreatment methods for biomass are typically combined with physical procedures including steam explosions, sulfur dioxide explosions, and ammonia fiber explosions (AFEX) (Badiei et al. 2014). According to Silverstein et al. (2007), RS hydrothermal or acid pretreatment resulted in more solubilization of hemicellulose and cellulose (Fig. 8.4).

8.3.6 Alkali Pretreatment

The amount of lignin was reduced by 36% after chopped RS was pretreated with alkali (20% solids and 2% NaOH loading for 1 h at 85 °C) (Zhang and Cai 2008). The split microfibrils improved the RS's exterior surface area and porosity, which made hydrolysis by enzymes easier (Tarkov and Feist 1969). According to Cheng et al. (2010), the ideal temperature for the reaction of RS pretreatment with NaOH and lime to remove about 23% to 27% lignin is 95 and 55 °C.

Wheat straw can be effectively pretreated by alkaline wet oxidation with Na_2CO_3, which yields solid fractions with 96% cellulose recovery (96%) and a 67% conversion rate to glucose by enzymes (Klinke et al. 2002). The complete procedure took place at a low chemical concentration (0.2–0.5%) and low temperature (25–40 °C), which led to lesser waste liquor production and comparatively inexpensive chemical costs. Following the two pretreatments, 90% of the cellulose that was originally present was recovered and lignin degraded by 81% (Curreli et al. 1997).

8.3.7 Ammonia Treatment

Ammonia has strong specificity for lignin reactions that cleave C-O-C bonds and ester and ether linkages in the complex of lignin and carbohydrate, rendering it a useful reagent for swelling of lignocellulosic materials (Kim and Lee 2007). Ammonia recycling percolation, a flow-through method, was created for pretreatment. A biomass bed kept at 170 °C is used for pumping ammonia during this operation. Using this method, it is possible to get up to delignification of 85% and a nearly theoretical yield of 75% of glucose during hydrolysis by enzymes (Drapcho et al. 2008). The majority of the samples' glucan and xylan were preserved by the pretreatment method, which involved immersing the samples in aqueous ammonia at temperatures between 40 and 90 °C for prolonged reaction times (Kim and Lee 2007; Kim et al. 2008). The highest enzymatic digestibility (71.1%) was obtained after pretreating RS with an aqueous ammonia solution (21% (w/w)) at medium temperatures of 69 °C for 10 h. One method to make xylitol and ethanol is by the following fermentation of hydrolysate from ammonia-treated RS (Swain and Chandraraj 2015).

Anhydrous ammonia is used in the AFEX procedure. No carbohydrate fraction is lost during this treatment, and there is no requirement to modify the pH (Ko et al. 2009). Biomass treated by AFEX generates up to 80% xylose theoretical yield and up to 90% theoretical yield of glucose when it is hydrolyzed using an enzyme. According to reports, AFEX is a successful pretreatment method for RS because pretreatment produced a loss of sugar of only 3% (Zhong et al. 2009). Ferrer et al. (1997) used an ammonia reactor unit of laboratory scale that consisted of a 4-L reactor together with the necessary additional apparatus to pretreat RS using a procedure named ammonia depressurization and pressurization. The amount of sugar production increased significantly as a result of pretreatment and enzymatic hydrolysis.

8.3.8 Acid Pretreatment

Because the technique cannot remove lignin and produces little sugar, there is a dearth of studies on the subject of hydrolysis by dilute acid of RS (Sumphanwanich et al. 2008). Wheat straw that has been pretreated with a weak acid (0.75%) yields better enzymatic saccharification. Enzyme detoxification and wheat straw hydrolysate treated with acid by over-liming decreased the time of fermentation to 39 h from 118 h in the case of SHF (pH 6.5, 35 °C), and as for SSF (pH 6.0, 35 °C), the yield of ethanol increased from to 17 g/L from 13 g/L and time of fermentation declined to 112 h from 136 h (Saha et al. 2005). Because it contains more cellulose than steam-exploded wheat straw alone does, it is a viable staple for ethanol manufacturing (Zabihi et al. 2010). By adjusting the period of pretreatment, temperature, and maleic acid concentration, Kootstra et al. (2009) stated the effects of wheat straw pretreatment with diluted maleic acid. The amount of furfural that formed was constrained, and almost all of it was converted to glucan and xylan through subsequent enzymatic hydrolysis. Wheat straws can be pretreated with diluted maleic or fumaric acid as a substitute for diluted sulfuric acid (Kootstra et al. 2009).

8.3.9 Oxidizing Agent Pretreatment

It is referred to as oxidative pretreatment when oxidizing compounds like peracetic acid or hydrogen peroxide are introduced to the biomass during specific pretreatment processes. Hemicellulose and lignin are removed during the pretreatment to make the cellulose more accessible. During oxidative pretreatment, significant events such electrophilic substitution, breaking of alkyl aryl ether linkages, side chain displacement, or oxidative cleavage of aromatic nuclei occur (Hon and Shiraishi 2001). The lignocellulosic matrix was loosened, and lignin was solubilized, increasing the enzyme digestibility despite no obvious alterations to the structure of the highly polymerized cellulose (Martel and Gould 1990). To describe how the lignocellulosic structure of straw has changed, Wei and Cheng (1985) examined fluctuations in accessibility for cadoxen solvent, the concentration of lignin, loss of weight, crystallinity of straw, and water-holding capacity. In comparison to untreated straw, pretreatment for 5 h at 60 °C in a 1% (w/w) solution of H_2O_2 and NaOH caused delignification by 60%, weight loss by 40%, a rise in Cadoxen accessibility by about five times, a rise in water-holding capacity by 11%, and just a modest drop in crystallinity. There are reports of using acetic acid to pretreat RS (Taniguchi et al. 1982). The cellulose's crystallinity in the straw had little to no degradation as a result of the acetic acid treatments. After treatment with 20% peracetic acid, the percentage of solubilization by enzymes in relation to the quantity of leftover straw was 42%. Alkaline peroxide pretreatment increased the number of monomeric sugars of wheat straw generated during hydrolysis by enzymes (8.6% (w/v)) (Saha and Cotta 2006). According to Jahan et al. (2006), RS was pretreated with either acetic acid or formic acid. Maximum pentosan dissolution was seen in an 80 °C, 0.6% H_2SO_4 catalyst, and 80% acetic acid solution for 120 min. Pentosan was broken down by acetic acid slower than by formate.

8.3.10 Organosolv Pretreatment

Delignification and hemicellulose removal are mostly handled by organosolv pretreatment, leaving a residue rich in cellulose that increases hydrolysis by enzymes at higher rates. Although the effect of organosolv pretreatment on cellulose's crystallinity is unknown, it has been shown that the kind of organic solvent used, as well as its concentration and temperature, all have a substantial impact on how much cellulose expands (Koo et al. 2011). To fraction the various components of biomass, the organosolv method utilizes heated organic solvents, like ethanol, that have an acidic pH. Organsolov pretreatment did not require the addition of any acidic material when carried out at high temperatures (185–210 °C) (Sun and Cheng 2002). RS was pulped using ethylene glycol and diethylene glycol combined with atmospheric pressure (Xuebing et al. 2009). Moreover, the breakdown of carbohydrates into undesirable furfural and hydroxymethylfurfural was decreased by a mild pretreatment temperature, pressure, and pH condition (Sun and Cheng 2002). Wheat straw was able to produce rational enzymatic hydrolysis products of the wet substrate (75%) and dried substrate (63%) thanks to the autocatalytic organosolv pretreatment-crude glycerol. The reasons for the low yield of enzymatic hydrolysis of dry pretreatment wheat straw were lignin-carbohydrate complex and/or lignin recondensation (Sun and Chen 2008a, 2008b). Based on the yield of fermentable sugar, Sindhu et al. (2010) indicated that RS pretreated with acetone can serve as a good source of raw materials for the manufacturing of bioethanol.

8.3.11 Pretreatment with Ionic Liquid

ILs are molten salts that contain organic or inorganic anions or organic cations. Because of their lack of flammability, high thermal and chemical stability, and low vapor pressure, ionic liquids were once regarded as "eco-friendly" (Hallett and Welton 2011). The lignocellulosic biomass undergoes considerable cellulose decrystallization and/or extensive lignin extraction during the IL pretreatment, which enhances the succeeding breakdown by enzymes (Lee et al. 2009). For biomass pretreatment, attempts to produce novel ILs from biomaterials that are renewable and natural have been made (Zhang et al. 2012). It is seen that the ionic liquid anions are primarily responsible for the breakdown of lignocellulose and cellulose (Wang et al. 2012). There has been less research on ILs that dissolve the lignin and the impact of structures of IL on lignin dissoluteness than there has been on cellulose-dissolving ILs (Hossain and Aldous 2012). Using glucose and pinoresinol as models for cellulose and lignin, respectively, Casas et al. (2012) investigated the cellulose and lignin dissolution in ionic liquids through simulation by computer and discovered that, similar to the dissolution of cellulose in ILs, the anion is crucial to the lignin dissolution process.

Pretreatment of RS with an ionic liquid, 1-ethyl-3-methylimidazolium acetate ([Emim][OAc]) before enzymatic hydrolysis, and yield of glucose was dramatically enhanced, going to more than 75% for the treated straw from 25.7% for the untreated straw, according to Poornejad et al. (2014). Similarly, cholinium lysine IL pretreatment considerably expanded the pore volume and surface area, which greatly increased the availability of polysaccharides to enzymes and improved the digestion of polysaccharides. Enzymatic hydrolysis of pretreated RS produced sugar yields of 48% for xylose and 81% for glucose. The hemicellulose and cellulose-rich fractions produced by pretreating wheat straw with 1-ethyl-3-methylimidazolium acetate ([Emim] [CH_3COO]) fully disintegrated biomass further demonstrated the extremely high degradability of the samples of cellulose and confirmed that the improved pretreatment technique produced a high glucose yield (da Costa Lopes et al. 2013).

8.3.12 Pretreatment by Biological Methods

White-rot fungi like *Phlebia subserialis*, *Ceriporiopsis subvermispora*, *Pleurotus ostreatus*, and *Phanerochaete chrysosporium* may metabolize lignin (Keller et al. 2003; Taniguchi et al. 2005). Many actinomycetes generate the enzymes required for the lignocellulose to be broken down in a manner similar to this. According to Zhang et al. (2016), RS was physiologically pretreated using *Streptomyces griseorubens* JSD-1. The *St. griseorubens* JSD-1 cellulolytic enzyme favored the conversion of cellulose to reducing sugars after 10 days of incubation, leading to an ideal saccharification efficiency of 88.13%. According to a 2011 article by Zeng et al., the addition of inorganic salts to wheat straw solid-state fermentation greatly increased the rate of biomass breakdown in *Ph. chrysosporium*. The digestibility of wheat straw peaked in 21 days after a subsequent pretreatment with ligninolytic fungus (*Irpex lacteus* and *Pleurotus eryngii*) after a moderate alkali treatment (López-Abelairas et al. 2013). For wheat straw pretreatment, five different fungi were identified by screening (Patel et al. 2006). The maximum yields of ethanol and total sugars were attained after *Aspergillus awamori* and *Aspergillus niger* pretreatment.

8.3.13 Combined Pretreatment

According to Kun et al. (2009), the processing of RS by alkaline material with the aid of photocatalysis effectively altered the material's physical characteristics and microstructure while lowering its lignin level. While the RS treatment with alkali with H_2O_2 exclusion favored the dissolution of the hemicelluloses with small molecular size, which are derived from α-glucan and rich in glucose, the second-stage treatment with alkaline peroxide improved the disintegration of xylose-rich hemicelluloses with large molecular size (Sun et al. 2002). Kim et al. (2013) described a two-step method of pretreatment of RS for the generation of fermentable sugar using aqueous ammonia and diluted sulfuric acid. The processing of RS with diluted aqueous ammonia and sulfuric acid resulted in fermentable sugar with a crystallinity index that was almost identical to that of cellulose. According to Zhu et al. (2006), RSs may be microwave processed in a number of methods to get rid of lignin and hemicelluloses. The effects of ILs and ultrasonic irradiation for pretreating RS under various ways were explored by Yang and Fang (2014). The greatest enzymatic efficiency among the hydrolyzed RS samples was achieved by ILs and ultrasonic pretreatment. Jin and Chen (2006) looked at how RS would react to steam explosion, ultrafine grinding, and enzymatic hydrolysis. Low-severity steam explosion and superfine grinding were combined for RS treatment in order to speed up the grinding process, save energy costs, avoid inhibitors, and boost enzymatic hydrolysis. To minimize excessive hemicellulose degradation and the generation of by-products from lignin and sugars, superfine grinding was done after the explosion of RS with steam at a minimal Ro (steam explosion severity factor). It revealed discrepancies between the residue of RS exploded with steam and steam exploded superfine ground RS product in terms of chemical compositions, enzymatic hydrolysis, properties of fiber, and contents of the cell.

8.4 Enzymatic Hydrolysis

The second stage in the generation of ethanol process from lignocellulosic substances is enzymatic hydrolysis. The cellulolytic enzyme is used to cleave the hemicellulose and cellulose polymers (Zhang and Lynd 2004). While cellulose normally only includes glucans, hemicellulose is made up of polymers of many sugars, including glucan, xylan, galactan, arabinan, and mannan. As a result, glucose is the main by-product of cellulose hydrolysis, while hemicellulose generates a variety of pentoses and hexoses (Taherzadeh and Niklasson 2004). However, the increased lignin concentration reduces yield and hydrolysis rate, restricts enzyme accessibility, and inhibits end products. Lignin, cellobiose, and glucose all work well to block cellulases (Knauf and Moniruzzaman 2004). Enzymatic hydrolysis of hemicellulose often takes place under mild conditions, such as a lengthy retention time and low pressure. After 48 h of enzyme hydrolysis, straw pretreatment with diluted sulfuric acid had a sugar yield of 0.72 g/g, which was higher than that of steam-pretreated straw (0.46 g/g) and treated straw (0.60 g/g) (Abedinifar et al. 2009). Aderemi et al.'s (2008) research looked at the kinetics of *A. niger*'s synthesis of glucose from RS. The research demonstrated that cell loading, substrate concentration, and RS pretreatment all had an impact on the rate and volume of glucose production. In contrast, RS treated with ammonia had an increase in monomeric sugars from 11% to 61% when compared to RS that hadn't been treated (Kun et al. 2009). Alkali was used to photocatalyze the hydrolysis of RS, and the enzyme approach produced 2.56 times more hydrolysis than the alkali method did (Sulbaran-de-Ferrer et al. 2003). Using a combination of enzymes, such as pectinases, xylanases, and cellulase rather than only cellulase, improves the efficiency of hydrolyzing lignocellulose biomass (Sulbaran-de-Ferrer et al. 2003).

After being enzymatically hydrolyzed for 12 h, the reducing sugars yield from wheat straw pretreatment reached 54.8% with these ILs at 130 °C for 30 min. After 48 h, the pretreated wheat straw with ambient aqueous glycerol autocatalytic organosolv showed a hydrolysis yield of above 90%. By eliminating the chemical barrier and changing the physical structural hindrance, this approach improved the saccharification of biomass containing lignocellulose (Sun and Cheng 2002). According to Zhu et al. (2006), microwave-assisted alkali pretreatment accelerated the rate of enzymatic hydrolysis as well as the amount of lignin and hemicellulose extracted from wheat straw. Nevertheless, the results of the second study revealed that physically crushing and irradiating wheat straw before pretreatment both enhanced saccharification (Yang et al. 2008).

8.5 Fermentation

An important measure in the manufacture of bioethanol is fermentation, which results in the direct generation of ethanol from the metabolism by the fermenting agent. Due to the complicated nature of the lignocellulosic hydrolysate, the procedure of fermentation is particularly crucial (Fig. 8.5, 8.6). Inhibitors, pentose, and hexose sugar created throughout the pretreatment of biomass make up the majority of the hydrolysate's chemical makeup. Hexose sugars can be easily fermented with ordinary yeast. Nevertheless, in order to use pentose sugar, a particular fermentation agent (yeast or bacteria) must be present, depending on how well it can digest pentose sugar. Theoretically, 0.51 kg of ethanol and 0.49 kg of carbon dioxide can be produced for every kilogram of glucose and xylose (Hamelinck et al. 2005). Temperature and a certain pH range are necessary for the bacteria employed in fermentation (Demirbas 2004). Since most fermentation agents are mesophilic organisms, they thrive in temperatures between 30 and 35 °C (Hettenhaus 1998).

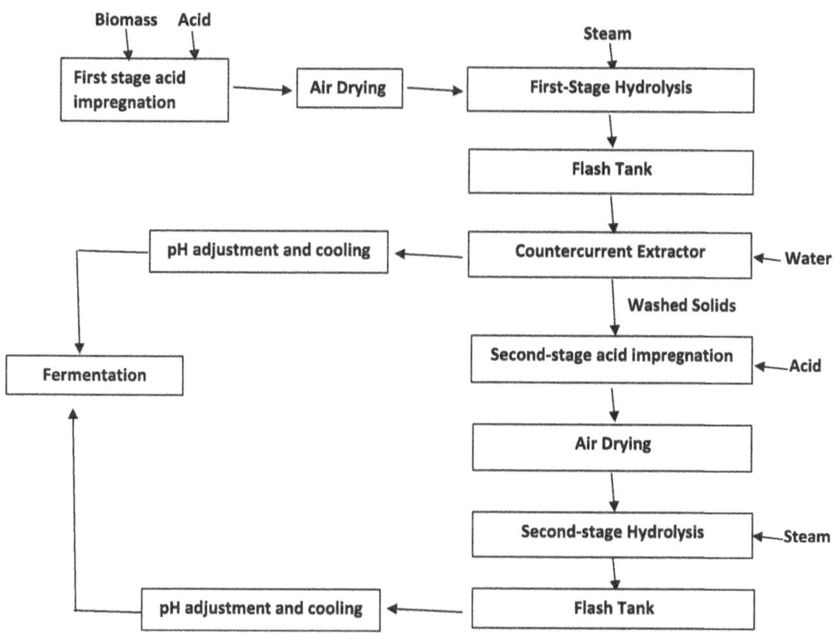

Fig. 8.5 Schematic flow diagram of the NREL's two-stage dilute sulfuric acid pretreatment process (Source: Drapcho et al. (2008)

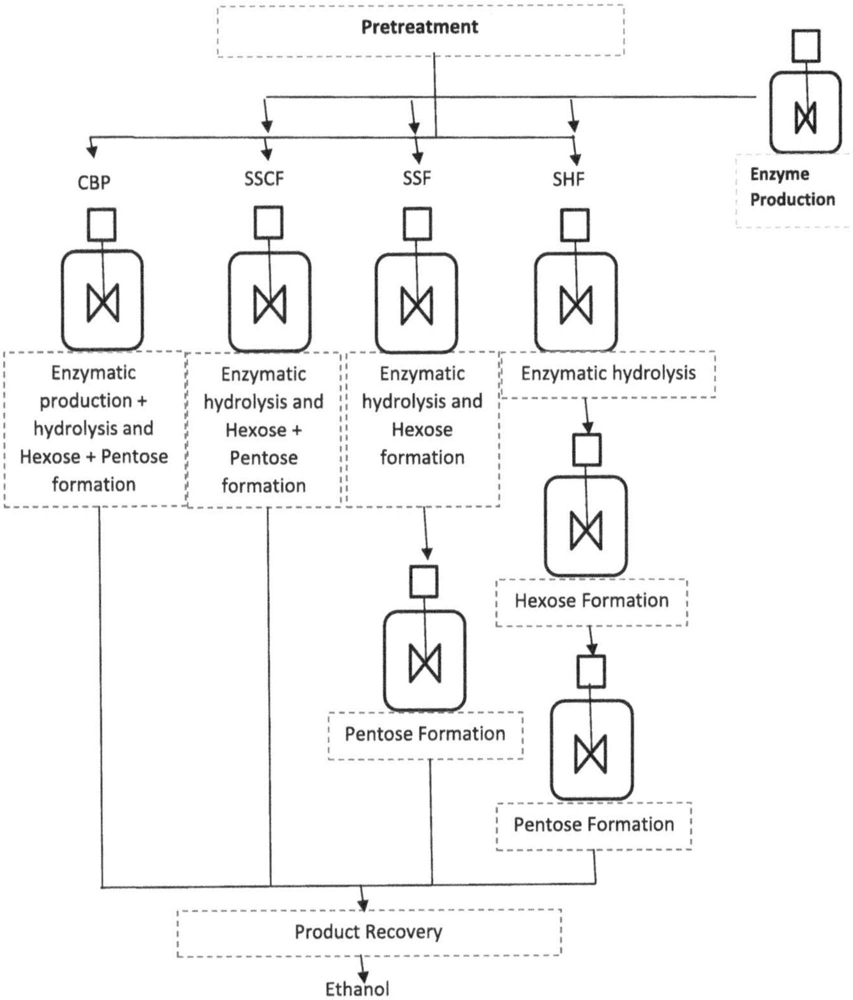

Fig. 8.6 Schematic diagrams of different available ethanol production methods from biomass of lignocellulose

pH values between 6.5 and 7.5 are normally required for fermentation germs to support growth (Aminifarshidmehr 1996). In comparison to bacteria, environments with a pH range of 3.5 to 5.0 are suitable for fungi to thrive in (Balat et al. 2008). The pace of tolerance to inhibitors, microbial growth and genetic stability, alcohol, osmosis, ethanol yield, and productivity are additional elements that are crucial to the fermentation process (Demirbas 2004). Many bacteria have been found that are suitable as fermentation microorganisms. Often employed as a fermenting bacterium, *Zymomonas mobilis* produces ethanol from glucose, sucrose, and fructose. Hexose sugars are naturally consumed by *Saccharomyces cerevisiae*, just like

Z. mobilis (e.g., glucose, fructose). Even though they are the most commonly used microorganisms, neither *S. cerevisiae* nor *Z. mobilis* can ferment pentose carbohydrates. The capacity of *Pachysolen tannophilus*, *Pichia stipitis*, and *Candida shehatae* for metabolization of xylose, a pentose sugar, is well-known. Several studies seek to genetically alter the behavior and function of bacteria and yeast due to their limitations as natural organisms. It is possible to give yeasts the ability to ferment additional sugar types through metabolic engineering. According to Buaban et al. (2010), after hydrolyzed sugarcane bagasse was fermented with engineered *Pi. stipitis* BCC15191 for 24 h acting as the medium for fermentation, 8.4 g/L of ethanol was produced. According to the study, the natural *Pi. stipitis* only had the ability to ferment xylose sugar, whereas the engineered *Pi. stipitis* BCC15191 had the capability to ferment glucose and xylose. Table 8.4 contains a list of the modified microorganisms used to increase ethanol output (Aditiya et al. 2016).

8.5.1 Separate Hydrolysis and Fermentation

This method provides a number of processing benefits. It allows fermentation enzymes to work at elevated temperatures for better execution while allowing fermenting organisms to work at mild temperatures, increasing sugar usage (Wingren et al. 2003).

8.5.2 Simultaneous Saccharification and Fermentation

Since the inhibitors for hydrolysis are removed by fermentation, SSF is said to produce a greater ethanol output, because of which this strategy is more appealing (Gupta and Verma 2015; Sun and Chen 2008a, 2008b). Watanabe et al. (2012) used the SSF method to extract ethanol from RS fronds, and it produced 38 g/L of ethanol or roughly 84.7% of the expected produce. Controlling the solution environment is difficult, though, because enzymes and microorganisms can only be used at the preferred temperature and pH. Although those conditions jeopardize the viability of microorganisms, most enzymes can still be made acceptable in mildly acidic environments (pH 5), and they can withstand temperatures of greater than 31 °C (Huang et al. 2005).

8.5.3 Simultaneous Saccharification and Co-fermentation

Although fermentation and hydrolysis occur simultaneously in the SSF process, the primary target for fermentation is hexose sugar, while pentose sugar is left unaffected. With the help of the SSCF method, mixed-culture microorganisms can begin

Table 8.4 List of engineered microorganisms used in bioethanol production improvement (Aditiya et al. 2016)

Employed microorganism	Microorganism type	Ethanol yield	Features
C. shehatae NCL3501	Yeast	Immobilized cells produced 0.5 g/g of sugar while autohydrolysis produced 0.45 g/g of sugar	Ferment xylose and glucose together
Clostridium thermocellum DSM1313	Bacterium	0.8 g/L	Increased ethanol production
Cl. thermocellum YD01	Bacterium	1.33 mol-ethanol/mole-glucose equivalents, or 3.0 times more than the wild type	Improved ethanol generation
Cl. thermocellum YD02	Bacterium	1.28 mol-ethanol/mole-glucose equivalent or three times as much as the wild type	Improved ethanol generation
Escherichia coli KO11	Bacterium	After a 48-h fermentation, 31.5 g/L, or potentially 91.5%	Ferment glucose and xylose
E. coli FBR5	Bacterium	2.25% (w/v)	The process of fermenting arabinose and xylose was bio-abated by *Coniochaeta ligniaria* NRRL30616
Pi. stipitis A	Yeast	0.4 g of product per gram of substrate	Bacterium concentration rose after it was hydrolyzed
Pi. stipitis NRRL Y-7124	Yeast	0.3 g of product per gram of substrate	Adapted at increasing concentration in hydrolysate
Pi. stipitis BCC15191	Yeast	8.4 g/L after a 24-h fermentation	Ferment glucose and xylose
S. cerevisiae D5a	Yeast	0.58% (w/v) or 100% theoretical yield	Enhanced yield of ethanol
S. cerevisiae RWB217	Yeast	0.43 g/g of sugars	Ferment xylose and glucose
Z. mobilis ZM4(pZB5)	Bacterium	28 g/L with the addition with 5 g/L yeast extract	Ferment glucose and xylose
Z. mobilis AX101	Bacterium	In the absence of acetic acid, 3.54 g/L/h; in the presence of acetic acid, 1.17 g/L/h	Ferment glucose, arabinose, and xylose
Thermoanaerobacterium saccharolyticum ALK2	Bacterium	37 g/L	Ethanol with a higher yield that can also ferment mannose, glucose, arabinose, and xylose
Thermoanaerobacter mathranii BG1L1	Bacterium	0.39–0.42 g/g sugars	Enhanced yield of ethanol

a continuous process without sugar separation and can use a variety of materials as substrates without having to sterilize them (Kleerebezem and van Loosdrecht 2007). For the manufacture of ethanol from wheat straw, RS, and other biomass, the SSCF process has been successfully tested (Moreno et al. 2013). According to Moreno et al. (2013), 20% solid loading of wheat straw which was laccase-mediated and pretreated with acid resulted in 22 g/L ethanol. Olofsson et al. (2010) stated a 35 g/L concentration of ethanol from wheat straw that had undergone steam pretreatment and an 11% solid loading. RS that had been mechanically milled and hydrothermally processed yielded ethanol (52 g/L) from solid loading (25%) in the SSCF (Sasaki et al. 2015).

8.6 Consolidated Bioprocessing (CBP)

CBP is one more alternative method. This method uses a cocktail of microbes to accomplish self-cellulase synthesis, substrate hydrolysis, and pentose and hexose fermentation all in one reactor for producing ethanol from cellulosic materials (van Zyl et al. 2007), comparing CBP to SSCFF and SSF.

The advantages of the approach include cheaper manufacturing costs due to fewer processes and no additional enzyme purchases, improved conversion efficiency, and less energy consumption for the manufacturing system (Carere et al. 2008). Although the use of specific microbes in the CBP approach has been described in various studies, it is uncommon for wild types of bacteria to do so. A recognized cellulolytic fungus, *Trichoderma reesei*, secretes cellulases for the breakdown of biomass cellulose (Xu et al. 2009). Moreover, a number of fungi, including *Fusarium oxysporum*, *Paecilomyces* sp., *Clostridium thermocellum*, and *Neurospora crassa*, are supposed to be useful for consolidate bioprocessing (Sarkar et al. 2012). Unfortunately, a slow ethanol conversion rate and a very low yield of ethanol are obstacles to the development of the CBP approach (Xu et al. 2009).

The only two categories into which the routes of modification in CBP may be classified are:

1. Changing cellulolytic microbe strains (cellulase producers) to also create ethanol (ethanologen). The potential of various fungi for this strain kind of change was described by Amore and Faraco (2012). These include *Aspergillus* species (primarily *Aspergillus niger*, *Aspergillus sojae*, *Aspergillus terreus*, and *Aspergillus oryzae*), T. *reesei*, *Rhizopus* species (*R. koji*, *R. oryzae*, and *R. stolonifer*), and *F. oxysporum*
2. Changing the genes of innately ethanol-producing bacteria to provide them with cellulolytic abilities. *Klebsiella oxytoca*, *Z. mobilis*, and *Escherichia coli* are only a few of the bacteria that have reportedly undergone modification. Yeasts transformed include *P. tannophilus*, *C. shehatae*, *Pi. stipitis*, and *S. cerevisiae* (Aditiya et al. 2016).

8.7 Analysis of RS Biorefinery's Bioeconomy

It is commonly acknowledged that waste biomass is a valuable supply that, when used well to develop the bio-based economy, could significantly replace petroleum feedstock. To fully analyze the rice-straw-based biorefineries value chain, it is essential to identify the researched system and its boundaries. The bulk of biorefineries based on biomass mainly focus on producing ethanol and other second-generation biofuels.

A biorefinery concept, on the other hand, is under development to boost sustainability and bio-economics by creating all conceivable useful goods from readily available waste biomass (Ranganathan 2022). The five steps for a biorefinery using RS separate fermentation and co-fermentation, liquefaction of lignin, synthesis of compounds from lignin, and the production of furfurals from hemicellulose are all processes carried out in a biorefinery. Furfurals are produced via separate fermentation and biochemical synthesis of hydrolyzed sugars. The separate hydrolysis and fermentation are useful for synthesizing of chemicals from cellulose (Bhatnagar et al. 2022; Harun et al. 2022). Co-fermentation has reportedly been shown to reduce a biofuel biorefinery's overall production cost (Demichelis et al. 2020). During biomass processing, the creation of co-products lowers expenses and improves the environmental friendliness of the procedure. Similar research revealed that the price of the entire process is reduced when bioethanol is produced with other chemical wastes. High-value compounds including ferulic acid, acetic acid, catechol, acetaldehyde, phenols, formic acid, and others are produced by this process (Bbosa et al. 2018). According to earlier studies by Jarunglumlert and Prommuak (2021) on techno-economics, co-production strategies help bring down the minimum sale cost of bioethanol close to market value, suggesting the likelihood of industrializing generation of bioethanol as a crucial alternative to fossil fuels. Although the RS burning in open fields harms the environment, it may also be used to make important goods like biocoal, which is used in power plants to generate bioelectricity. At first, RS was utilized to produce bioethanol, but this idea fell short of being a sustainable and economical biorefinery. Instead of using it for only one product, RS should be channeled to several products (Zhou et al. 2021). A cost-effective no-waste RS biorefinery can therefore be developed by extrapolating from the available literature that technology developments can help reduce overall processing costs, provide high-value co-products, and ease environmental constraints. By establishing a waste biomass-based bioeconomy and producing a range of goods, we can manage waste, create wealth, protect the environment, and open up several opportunities for jobs.

8.8 The Difficulties with Biorefining RS

Researchers have offered a variety of environmentally responsible and sustainable methods for turning RS waste into beneficial goods through green biorefinery (Jusoh et al. 2013; Ma et al. 2022; Saritha et al. 2013; Sweeney and McCouch 2007). Yet, some GHGs that can be turned into commercial CNG are still stimulated and worsened during the processing of biomass employing recent technical advancements:

1. The chemical and thermal processing of RS results in significant volumes of ash and caustic alkali metals. Moreover, the production of slag in the combustion has the potential to damage boilers and heaters, impairing their overall performance. However, this issue can be resolved, and the expense of the process can be made up by using this slag and ash as cementing material (Bhatia et al. 2021; Zaky et al. 2008).
2. As insects destroy the straw and alter the raw material quality, the harvesting time also influences the distinctive composition of the raw material. To procure high-quality biochemicals and value-added products, and to ensure that the material used for different technological procedures has a similar chemical makeup, this issue can be overcome by establishing a suitable structure for harvesting on time (Saritha et al. 2013).
3. The ongoing usage of chemical fertilizers has an impact on the RS's ingredient composition as well. Fertilizers, biofertilizers, and organic manure can be used in place of RS compost to increase the RS and crop quality, cut back on the usage of pricey chemical fertilizers, provide humus to the soil, and prevent pest-and insect-borne diseases (Abo-State et al. 2014; Aysu and Kucuk 2014).
4. Thirty to forty percent of the overall production cost is made up of the raw materials and their logistics value chain. To lower logistics costs, a method must be designed for the appropriate collection, processing (chopping, drying), and raw material transportation. According to earlier researchers, the feedstock needs to be compressed into minor pallets that can be shipped readily and can aid in minimizing costs for a feasible biorefinery (Azadi et al. 2013).
5. Drying is a crucial component of biomass processing in biorefineries, particularly when bioenergy is being extracted from biomass. Yet, it may be advantageous to eliminate the moisture content at a reduced cost by using solar radiation and its associated natural renewable technologies. Transportation, time, and other logistics expenses can be decreased when raw materials are transported after drying (Erdocia et al. 2014).
6. For RS, using compositional extraction techniques employing ultra-critical (CO_2) processes and other clean technologies comes at a high commercial cost. To sustainably control the total expenditures of this green biorefinery, these technologies will, however, result in compositional fractions that are superior in quantity and quality. Further, using cutting-edge biorefinery methods and scientific interventions, these fractions can be converted into any feasible goods.

8.9 Conclusion

In conclusion, the production of bioethanol from lignocellulosic waste derived from paddy straw presents a sustainable approach to the provision of renewable energy and trash management. Through improvements in pretreatment techniques, enzyme technologies, and fermentation techniques, ongoing research seeks to optimize the bioethanol production process, increase efficiency, and lower prices. Paddy straw lignocellulosic waste may be used to produce bioethanol, which not only suggests a different energy source but also lowers greenhouse gas emissions and lessens agricultural waste. The discovery and application of this technology might open up new doors for rural areas, fostering economic development and job creation. In general, producing bioethanol from lignocellulosic waste like paddy straw is a realistic and ecologically responsible way to tackle our current energy and waste management problems.

References

Abedinifar S, Karimi K, Khanahmadi M, Taherzadeh MJ (2009) Ethanol production by *Mucor indicus* and *Rhizopus oryzae* from rice straw by separate hydrolysis and fermentation. Biomass Bioenergy 33:828–833

Abo-State MA, Ragab AM, Gendy NS, Farhat LA, Maidan HR (2014) Bioethanol production from rice straw enzymatically saccharified by fungal isolates, Trichoderma viride F94 and Aspergillus terreus F98. Sci Res 3:19–29

Aderemi BO, Abu E, Highina BK (2008) The kinetics of glucose production from rice straw by *Aspergillus Niger*. Afr J Biotechnol 7:1745–1752

Aditiya HB, Mahlia TMI, Chong WT, Nur H, Sebayang AH (2016) Second generation bioethanol production: a critical review. Renew Sust Energ Rev 66:631–653

Aminifarshidmehr N (1996) The management of chronic suppurative otitis media with acid media solution. Am J Otolaryngol 17:24–25

Amiri H, Karimi K, Zilouei H (2014) Organosolv pretreatment of rice straw for efficient acetone, butanol, and ethanol production. Bioresour Technol 152:450–456

Amore A, Faraco V (2012) Potential of fungi as category I consolidated bioprocessing organisms for cellulosic ethanol production. Renew Sust Energ Rev 16:3286–3301

Aysu T, Kucuk MM (2014) Biomass pyrolysis in a fixed-bed reactor: effects of pyrolysis parameters on product yields and characterization of products. Energy 64:1002–1025

Azadi P, Inderwildi OR, Farnood R, King DA (2013) Liquid fuels, hydrogen and chemicals from lignin: a critical review. Renew Sust Energ Rev 21:506–523

Badiei M, Asim N, Jahim JM, Sopian K (2014) Comparison of chemical pretreatment methods for cellulosic biomass. APCBEE Proc 9:170–174

Bak JS (2014) Electron beam irradiation enhances the digestibility and fermentation yield of water-soaked lignocellulosic biomass. Biotechnol Rep 4:30–33

Bak JS, Ko JK, Han YH, Lee BC, Choi IG, Kim KH (2009) Improved enzymatic hydrolysis yield of rice straw using electron beam irradiation pretreatment. Bioresour Technol 100:1285–1290

Balat M, Balat H, Oz C (2008) Progress in bio-ethanol processing. Prog Energy Combust Sci 34: 551–573

Baxter LL, Miles TR, Miles TR Jr, Jenkins BM, Dayton DC, Milne TA, Bryers RW, Oden LL (1996) The behavior of inorganic material in biomass-fired power boilers—field and laboratory experiences: volume II of alkali deposits found in biomass power plants. National Renewable Energy Laboratory, Golden, CO

Bbosa D, Mba-Wright M, Brown RC (2018) More than ethanol: a techno-economic analysis of a corn stover-ethanol biorefinery integrated with a hydrothermal liquefaction process to convert lignin into biochemicals. Biofuels Bioprod Biorefin 12:497–509

Bhatia SK, Jagtap SS, Bedekar AA, Bhatia RK, Karthik R, Pugazhendhi A, Rao VC, Atabani AE, Kumar G, Yung-Hun Y (2021) Renewable biohydrogen production from lignocellulosic

biomass using fermentation and integration of systems with other energy generation technologies. Sci Total Environ 765:144429
Bhatnagar A, Khatri P, Krzywonos M, Tolvanen H, Konttinen J (2022) Techno-economic and environmental assessment of decentralized pyrolysis for crop residue management: rice and wheat cultivation system in India. J Clean Prod 367:132998
Binod P, Sindhu R, Singhania RR, Vikram S, Devi L, Nagalakshmi S, Kurien N, Sukumaran RK, Pandey A (2010) Bioethanol production from rice straw: an overview. Bioresour Technol 101: 4767–4774
Bouchard J, Méthot M, Jordan B (2006) The effects of ionizing radiation on the cellulose of wood free paper. Cellulose 13:601–610
Buaban B, Inoue H, Yano S, Tanapongpipat S, Ruanglek V, Champreda V, Pichyangkura R, Rengpipat S, Eurwilaichitr L (2010) Bioethanol production from ball milled bagasse using an on-site produced fungal enzyme cocktail and xylose-fermenting *Pichia stipitis*. J Biosci Bioeng 110:18–25
Callaway E (2014) Domestication. The birth of rice. Nature 514:S58–S59
Carere C, Sparling R, Cicek N, Levin D (2008) Third generation biofuels via direct cellulose fermentation. Int J Mol Sci 9:1342–1360
Casas A, Palomar J, Alonso MV, Oliet M, Omar S, Rodriguez F (2012) Comparison of lignin and cellulose solubilities in ionic liquids by COSMO-RS analysis and experimental validation. Ind Crop Prod 37:155–163
Chang VS, Holtzapple MT (2000) Fundamental factors affecting biomass enzymatic reactivity. Appl Biochem Biotechnol 84–86(1):5–37
Cheng YS, Zheng Y, Yu CW, Dooley TM, Jenkins BM, Vander Gheynst JS (2010) Evaluation of high solids alkaline pretreatment of rice straw. Appl Biochem Biotechnol 162:1768
Curreli N, Fadda MB, Rescigno A, Rinaldi AC, Soddu G, Sollai F, Vaccargiu S, Sanjust E, Rinaldi A (1997) Mild alkaline/oxidative pretreatment of wheat straw. Process Biochem 32:665–670
da Costa Lopes AM, Joao KG, Rubik DF, Łukasik EB, Duarte LC, Andreaus J, Bogel-Łukasik R (2013) Pre-treatment of lignocellulosic biomass using ionic liquids: wheat straw fractionation. Bioresour Technol 142:198–208
Delgenés JP, Penaud V, Moletta R (2002) Pretreatments for the enhancement of anaerobic digestion of solid wastes. In: Mata-Alvarez J (ed) Biomethanization of the organic fraction of municipal solid wastes. IWA Publishing, London, pp 201–228
Demichelis F, Laghezza M, Chiappero M, Fiore S (2020) Technical, economic and environmental assessment of bioethanol biorefinery from waste biomass. J Clean Prod 277:124111
Demirbas A (2004) Ethanol from cellulosic biomass resources. J Green Energy 1:79–87
Drapcho CM, Nhuan NP, Walker TH (2008) Biofuels engineering process technology. McGraw Hill, New York
Erdocia X, Prado R, Corcuera MÁ, Labidi J (2014) Base catalyzed depolymerization of lignin: influence of organosolv lignin nature. Biomass Bioenergy 66:379–386
Ferrer S et al (1997) Sugar production from rice straw. Arch Latinoam Prod Anim 5:112–114
Garrote G, Dominguez H, Parajo JC (2002) Autohydrolysis of corncob: study of non-isothermal operation for xylooligosaccharide production. J Food Eng 52:211–218
Gong G, Liu D, Huang Y (2010) Microwave-assisted organic acid pretreatment for enzymatic hydrolysis of rice straw. Biosyst Eng 107:67–73
Gupta A, Verma JP (2015) Sustainable bio-ethanol production from agro-residues: a review. Renew Sust Energ Rev 41:550–567
Hallett JP, Welton T (2011) Room-temperature ionic liquids: solvents for synthesis and catalysis. Chem Rev 111(5):3508–3576
Hamelinck C, Hooijdonk G, Faaij A (2005) Ethanol from lignocellulosic biomass: techno economic performance in short, middle and long term. Biomass Bioenergy 28:384–410
Harun SN, Hanafiah MM, Noor NM (2022) Rice straw utilisation for bioenergy production: a brief overview. Energies 15:5542

Hettenhaus J (1998) Ethanol fermentation strains: present and future requirements for biomass to ethanol commercialization. US Department of Energy and National Renewable Energy Laboratory, Golden, CO
Hideno A, Inoue H, Tsukahara K, Fujimoto S, Minowa T, Inoue S, Endo T, Sawayama S (2009) Wet disk milling pretreatment without sulfuric acid for enzymatic hydrolysis of rice straw. Bioresour Technol 100(10):2706–2711
Hon DNS, Shiraishi N (2001) Wood and cellulose chemistry, 2nd edn. Marcel Dekker, New York
Hossain MM, Aldous L (2012) Ionic liquids for lignin processing: dissolution, isolation, and conversion. Aust J Chem 65(11):1465–1477
Hsu TC, Guo GL, Chen WH, Hwang WS (2010) Effect of dilute acid pretreatment of rice straw on structural properties and enzymatic hydrolysis. Bioresour Technol 101:4907–4913
Huang L, Jin B, Lant P, Zhou J (2005) Simultaneous saccharification and fermentation of potato starch waste water to lactic acid by *Rhizopus oryzae* and *Rhizopus arrhizus*. Biochem Eng J 23: 265–276
Idris M, Hashim H (2021) Integrating palm oil biomass waste utilisation in coal-fired power plants for meeting near-term emission targets. J Environ Manag 296:113118
Jahan MS, Lee ZZ, Jin Y (2006) Organic acid pulping of rice straw. I: cooking. Turk J Agric For 30: 231–239
Janker-Obermeier I, Sieber V, Faulstich M, Schieder D (2012) Solubilization of hemicellulose and lignin from wheat straw through microwave-assisted alkali treatment. Ind Crop Prod 39:198–203
Jarunglumlert T, Prommuak C (2021) Net energy analysis and techno-economic assessment of co-production of bioethanol and biogas from cellulosic biomass. Fermentation 7:229
Jenkins BM, Baxter LL, Miles TR Jr, Miles TR (1998) Combustion properties of biomass. Fuel Process Technol 54:17–46
Jin S, Chen H (2006) Superfine grinding of steam-exploded rice straw and its enzymatic hydrolysis. Biochem Eng J 30:225–230
Jin SB, Ja KK, Young HH, Byung CL, In-Geol C, Heon KK (2009) Improved enzymatic hydrolysis yield of rice straw using electron beam irradiation pretreatment. Bioresour Technol 100:1285–1290
Jusoh ML, Manaf LA, Latiff PA (2013) Composting of rice straw with effective microorganisms (EM) and its influence on compost quality. Iran J Environ Health Sci Eng 10:17
Keller FA, Hamilton JE, Nguyen QA (2003) Microbial pretreatment of biomass: potential for reducing severity of thermo-chemical biomass pretreatment. Appl Biochem Biotechnol 105–108:27–41
Kim TH, Lee YY (2007) Pretreatment of corn stover by soaking in aqueous ammonia at moderate temperatures. Appl Biochem Biotechnol 137–140(1–12):81–92
Kim TH, Taylor F, Hicks KB (2008) Bioethanol production from barley hull using SAA (soaking in aqueous ammonia) pretreatment. Bioresour Technol 99:5694–5702
Kim SB, Lee SJ, Lee JH, Jung YR, Thapa LP, Kim JS, Um Y, Park C, Kim SW (2013) Pretreatment of rice straw with combined process using dilute sulfuric acid and aqueous ammonia. Biotechnol Biofuels 6:109
Kitchaiya P, Intanakul P, Krairish M (2003) Enhancement of enzymatic hydrolysis of lignocellulosic wastes by microwave pretreatment under atmospheric pressure. J Wood Chem Technol 23(2):217–225
Kleerebezem R, van Loosdrecht MCM (2007) Mixed culture biotechnology for bioenergy production. Curr Opin Biotechnol 18:207–212
Klinke HB, Ahring BK, Schmidt AS, Thomsen AB (2002) Characterization of degradation products from alkaline wet oxidation of wheat straw. Bioresour Technol 82:15–26
Knauf M, Moniruzzaman M (2004) Lignocellulosic biomass processing: a perspective. Int Sugar J 106:147–150

Ko JK, Bak JS, Jung MW, Lee HJ, Choi IG, Kim TH, Kim KH (2009) Ethanol production from rice straw using optimized aqueous-ammonia soaking pretreatment and simultaneous saccharification and fermentation processes. Bioresour Technol 100:4374–4380

Koga N, Tajima R (2011) Assessing energy efficiencies and green house gas emissions under bioethanol-oriented paddy rice production in northern Japan. J Environ Manag 92:967–973

Koo BW, Kim HY, Park N, Lee SM, Yeo H, Choi IG (2011) Organosolv pretreatment of *Liriodendron tulipifera* and simultaneous saccharification and fermentation for bioethanol production. Biomass Bioenergy 35:1833–1840

Kootstra AMJ, Beeftink HH, Scott EL, Sanders JPM (2009) Optimization of the dilute maleic acid pretreatment of wheat straw. Biotechnol Biofuels 2:31

Kun N, Peng C, Xu Z, Wen-Song T (2009) Enhanced enzymatic hydrolysis of rice straw pretreated by alkali assisted with photocatalysis technology. J Chem Technol Biotechnol 84:1240–1245

Laureano-Perez L, Teymouri F, Alizadeh H, Dale BE (2005) Understanding factors that limit enzymatic hydrolysis of biomass. Appl Biochem Biotechnol 124:1081–1099

Lee JW, Rodrigues RC, Jeffries TW (2009) Simultaneous saccharification and ethanol fermentation of oxalic acid pretreated corncob assessed with response surface methodology. Bioresour Technol 100:6307–6311

Lewis NG, Yamamoto E (1990) Lignin: occurrence, biogenesis and biodegradation. Annu Rev Plant Biol 41:455–496

López-Abelairas M, Álvarez Pallín M, Salvachúa D, Lú-Chau T, Martínez MJ, Lema JM (2013) Optimization of the biological pretreatment of wheat straw with white-rot fungi for ethanol production. Bioprocess Biosyst Eng 36(9):1251–1260

Ma H, Beadham I, Ruan W, Zhang C, Deng Y (2022) Enhancing rice straw compost with an amino acid-derived ionic liquid as additive. Bioresour Technol 345:126387

Maiorella BI (1983) Ethanol industrial chemicals. Biochem Fuels:861–914

Maiorella BL (1985) Ethanol. In: Moo-Young M (ed) Comprehensive biotechnology. Pergamon Press, Oxford, pp 861–914

Martel P, Gould JM (1990) Cellulose stability and delignification after alkaline hydrogen peroxide treatment of straw. J Appl Polym Sci 39:707–714

Matsumura Y, Minowa T, Yamamoto H (2005) Amount, availability and potential use of rice straw biomass as an energy resource in Japan. Biomass Bioenergy 29:347–354

Moreno AD, Tomás-Pejó E, Ibarra D, Ballesteros M, Olsson L (2013) In situ laccase treatment enhances the fermentability of steam-exploded wheat straw in SSCF processes at high dry matter consistencies. Bioresour Technol 143:337–343

Mussatto SI, Roberto IC (2004) Optimal experimental condition for hemicellulosic hydrolyzate treatment with activated charcoal for xylitol production. Biotechnol Prog 20:134–139

Nguyen TA, Kim KR, Han SJ, Cho HY, Kim JW, Park SM, Park JC, Sim SJ (2010) Pretreatment of rice straw with ammonia and ionic liquid for lignocellulose conversion to fermentable sugars. Bioresour Technol 101(19):7432–7438

Oberoi HS, Vadlania PV, Brijwania K, Bhargavb VK, Patil RT (2010) Enhanced ethanol production via fermentation of rice straw with hydrolysate-adapted *Candida tropicalis* ATCC 13803. Process Biochem 45:1299–1306

Oberoi HS, Babbar N, Sandhu SK, Dhaliwal SS, Kaur U, Chadha BS, Bhargav VK (2012) Ethanol production from alkali-treated rice straw via simultaneous saccharification and fermentation using newly isolated thermotolerant *Pichia kudriavzevii* HOP-1. J Ind Microbiol Biotechnol 39(4):557–566

Olofsson K, Palmqvist B, Lidén G (2010) Improving simultaneous saccharification and co-fermentation of pretreated wheat straw using both enzyme and substrate feeding. Biotechnol Biofuels 3:17

Palmowski L, Muller J (1999) Influence of the size reduction of organic waste on their anaerobic digestion. In: Proceedings of the II international symposium on anaerobic digestion of solid waste, Barcelona, 15–17 June, pp 137–144

Park J, Shiroma R, Haq M, Zhang Y, Ike M, Sanoh YA, Ida A, Kondo M, Tokuyasu K (2010) A novel lime pretreatment for subsequent bioethanol production from rice straw—calcium capturing by carbonation (CaCCO) process. Bioresour Technol 101(17):6805–6811
Patel SJ, Onkarappa R, Shobha K (2006) Study of ethanol production from fungal pretreated wheat and rice straw. Int J Microbiol 4:1–6
Poornejad N, Karimi K, Behzad T (2014) Ionic liquid pretreatment of rice straw to enhance saccharification and bioethanol production. J Biomass Biofuel 1:8–15
Ranganathan P (2022) Preliminary techno-economic evaluation of 2G ethanol production with co-products from rice straw. Biomass Conv Bioref 12:3673–3686
Rathour RK, Devi M, Dahiya P, Sharma N, Kaushik N, Kumari D, Kumar P, Baadhe RR, Walia A, Bhatt AK (2023) Recent Trends, Opportunities and Challenges in Sustainable Management of Rice Straw Waste Biomass for Green Biorefinery. Energies 16:1429.
Roberto IC, Mussatto SI, Rodrigues RCLB (2003) Dilute-acid hydrolysis for optimization of xylose recovery from rice straw in a semi-pilot reactor. Ind Crop Prod 7:171–176
Saha BC (2003) Hemicellulose bioconversion. J Ind Microbiol Biotechnol 30:279–291
Saha BC, Cotta MA (2006) Ethanol production from alkaline peroxide pretreated enzymatically saccharified wheat straw. Biotechnol Prog 22:449–453
Saha BC, Iten LB, Cotta MA, Wu YV (2005) Dilute acid pretreatment, enzymatic saccharification and fermentation of wheat straw to ethanol. Process Biochem 40:3693–3700
Saritha M, Arora A, Singh S, Nain L (2013) Streptomyces griseorubens mediated delignification of paddy straw for improved enzymatic saccharification yields. Bioresour Technol 135:12–17
Sarkar N, Ghosh S, Bannerjee S, Aikat K (2012) Bioethanol production from agricultural wastes: an overview. Renew Energy 37:19–27
Sasaki K, Tsuge Y, Sasaki D, Teramura H, Inokuma K, Hasunuma T, Ogino C, Kondo A (2015) Mechanical milling and membrane separation for increased ethanol production during simultaneous saccharification and co-fermentation of rice straw by xylose-fermenting *Saccharomyces cerevisiae*. Bioresour Technol 185:263–268
Sheehan J, Aden A, Paustian K, Killian K, Walsh M, Nelson R (2003) Energy and environmental aspects of using corn stover for fuel ethanol. J Ind Ecol 7:117–146
Silverstein RA, Chen Y, Sharma-Shivappa RR, Boyette MD, Osborne J (2007) A comparison of chemical pretreatment methods for improving saccharification of cotton stalks. Bioresour Technol 98:3000–3011
Sindhu R, Binod P, Nagalakshmi S, Janu KU, Sajna KV, Kurien N, Sukumaran RK, Pandey A (2010) Formic acid as a potential pretreatment agent for conversion of sugarcane bagasse to ethanol. Appl Biochem Biotechnol 162:2313–2323
Singh A, Bishnoi NR (2012) Enzymatic hydrolysis optimization of microwave alkali pretreated wheat straw and ethanol production by yeast. Bioresour Technol 108:94–101
Singh R, Srivastava M, Shukla A (2016) Environmental sustainability of bio-ethanol production from rice straw in India: a review. Renew Sust Energ Rev 54:202–216
Song Z, Yang G, Guo Y, Zhang T (2012) Comparison of two chemical pretreatments of rice straw for biogas production by anaerobic digestion. Bioresources 7:3223–3236
Stepanik TM, Rajagopal S, Ewing D, Whitehouse R (1998) Electron processing technology: a promising application for the viscose industry. Radiat Phys Chem 52:505–509
Sulbaran-de-Ferrer B, Aristiguieta M, Dale BE, Ferrer A, Ojeda-de-Rodriguez G (2003) Enzymatic hydrolysis of ammonia-treated rice straw. Appl Biochem Biotechnol 105–108:155–164
Sumphanwanich J, Leepipatpiboon N, Srinorakutara T, Akaracharanya A (2008) Evaluation of dilute-acid pre-treated bagasse corn cob and rice straw for ethanol fermentation by *Saccharomyces cerevisiae*. Ann Microbiol 58:219–225
Sun F, Chen H (2008a) Enhanced enzymatic hydrolysis of wheat straw by aqueous glycerol pretreatment. Bioresour Technol 99:6156–6161
Sun F, Chen H (2008b) Organosolv pretreatment by crude glycerol from oleochemicals industry for enzymatic hydrolysis of wheat straw. Bioresour Technol 99:5474–5479

Sun Y, Cheng J (2002) Hydrolysis of lignocellulosic materials for ethanol production: a review. Bioresour Technol 83:1–11

Sun RC, Tomkinson J, Ma PL, Liang SF (2002) Comparative study of hemicelluloses from rice straw by alkali and hydrogen peroxide treatments. Carbohydr Polym 42:111–122

Swain MR, Chandraraj K (2015) Improved conversion of rice straw to ethanol and xylitol by combination of moderate temperature ammonia pretreatment and sequential fermentation using *Candida tropicalis*. Ind Crop Prod 77:1039–1046

Sweeney M, McCouch S (2007) The complex history of the domestication of rice. Ann Bot 100: 951–957

Taherzadeh MJ, Niklasson C (2004) Ethanol from lignocellulosic materials: pretreatment, acid and enzymatic hydrolyses, and fermentation. ACS Symp Ser 889:49–68

Takano M, Hoshino K (2018) Bioethanol production from rice straw by simultaneous saccharification and fermentation with statistical optimized cellulase cocktail and fermenting fungus. Bioresour Bioprocess 5:16

Takkellapati S, Li T, Gonzalez MA (2018) An overview of biorefinery-derived platform chemicals from a cellulose and hemicellulose biorefinery. Clean Techn Environ Policy 20:1615–1630

Taniguchi M, Tanaka M, Matsuno R, Kamikubo T (1982) Evaluation of chemical pretreatment for enzymatic solubilization of rice straw. Eur J Appl Microbiol Biotechnol 14:35–39

Taniguchi M, Suzuki H, Watanabe D, Sakai K, Hoshino K, Tanaka T (2005) Evaluation of pretreatment with *Pleurotus ostreatus* for enzymatic hydrolysis of rice straw. J Biosci Bioeng 100(6):637–643

Tarkov H, Feist WC (1969) A mechanism for improving the digestibility of lignocellulosic materials with dilute alkali and liquid ammonia. Adv Chem 95:197–218

van Zyl W, Lynd L, Den Haan R, Mc Bride J (2007) Consolidated bio-processing for bioethanol production using Saccharomyces cerevisiae. Adv Biochem Eng Biotechnol 108:205–235

Wang CZ, He H, Wang X, Yuan CS (2012) Trends in scientific publications of Chinese medicine. Am J Chin Med 40:1099–1108

Watanabe I, Miyata N, Ando A, Shiroma R, Tokuyasu K, Nakamura T (2012) Ethanol production by repeated-batch simultaneous saccharification and fermentation (SSF) of alkali-treated rice straw using immobilized *Saccharomyces cerevisiae* cells. Bioresour Technol 123:695–698

Wei CJ, Cheng CY (1985) Effect of hydrogen peroxide pretreatment on the structural features and the enzymatic hydrolysis of rice straw. Biotechnol Bioeng 27:1418–1426

Wingren A, Galbe M, Zacchi G (2003) Techno-economic evaluation of producing ethanol from softwood: comparison of SSF and SHF and identification of bottlenecks. Biotechnol Prog 19(4): 1109–1117

Xiong J, Ye J, Liang WZ, Fan PM (2000) Influence of microwave on the ultrastructure of cellulose I. J South China Union Technol 28(3):84–89

Xu Q, Singh A, Himmel M (2009) Perspectives and new directions for the production of bioethanol using consolidated bioprocessing of lignocellulose. Curr Opin Biotechnol 20:364–371

Xuebing Z, Keke C, Dehua L (2009) Organosolv pretreatment of lignocellulosic biomass for enzymatic hydrolysis. Appl Microbiol Biotechnol 82:815–827

Yang CY, Fang TJ (2014) Combination of ultrasonic irradiation with ionic liquid pretreatment for enzymatic hydrolysis of rice straw. Bioresour Technol 164:198–202

Yang C, Shen Z, Yu G, Wang J (2008) Effect and after effect of γ radiation pretreatment on enzymatic hydrolysis of wheat straw. Bioresour Technol 99:6240–6245

Zabihi S, Alinia R, Esmaeilzadeh F, Kalajahi JF (2010) Pretreatment of wheat straw using steam, steam/acetic acid and steam/ethanol and its enzymatic hydrolysis for sugar production. Biosyst Eng 105:288–297

Zaky RR, Hessien MM, El-Midany AA, Khedr MH, Abdel-Aal EA, El-Barawy KA (2008) Preparation of silica nanoparticles from semi-burned rice straw ash. Powder Technol 185:31–35

Zamora R, Crispin JAS (1995) Production of an acid extract of rice straw. Acta Cient Venez 46: 135–139

Zeng J, Singh D, Chen S (2011) Biological pretreatment of wheat straw by *Phanerochaete chrysosporium* supplemented with inorganic salts. Bioresour Technol 102:3206–3214

Zevenhoven M (2000) The prediction of deposit formation in combustion and gasification of biomass fuels—fractionation and thermodynamic multi-phase multi-component equilibrium (TPCE) calculations. In: Haefele S (ed) Combustion and materials chemistry. Lemminkäinengatan, Turku, p 38

Zhang QZ, Cai WM (2008) Enzymatic hydrolysis of alkali-pretreated rice straw by *Trichoderma reesei* ZM4-F3. Biomass Bioenergy 32:1130–1135

Zhang YH, Lynd LR (2004) Toward an aggregated understanding of enzymatic hydrolysis of cellulose: noncomplexed cellulase systems. Biotechnol Bioeng 88(7):797–824

Zhang J, Tang M, Viikari L (2012) Xylans inhibit enzymatic hydrolysis of lignocellulosic materials by cellulases. Bioresour Technol 121:8–12

Zhang D, Luo Y, Chu S, Zhi Y, Wang B, Zhou P (2016) Biological pretreatment of rice straw with *Streptomyces griseorubens* JSD-1 and its optimized production of cellulase and xylanase for improved enzymatic saccharification efficiency. Prep Biochem Biotechnol 17:575–585

Zhong C, Lau MW, Balan V, Dale BE, Yuan YJ (2009) Optimization of enzymatic hydrolysis and ethanol fermentation from AFEX-treated rice straw. Appl Microbiol Biotechnol 84:667–676

Zhou Y, Searle S, Anup S (2021) Techno-economic analysis of cellulosic ethanol in India using agricultural residues—a report. International Council on Clean Transportation, Washington, DC

Zhu S, Wu Y, Yu Z, Liao J, Zhang Y (2005a) Pretreatment by microwave/alkali of rice straw and its enzymic hydrolysis. Process Biochem 40:3082–3086

Zhu S, Wu Y, Yu Z, Wang C, Yu F, Jin S, Ding Y, Chi RA, Liao J, Zhang Y (2005b) Comparison of three microwave/chemical pretreatment processes for enzymatic hydrolysis of rice straw. Biosyst Eng 93:279–283

Zhu S, Wu Y, Yu Z, Zhang X, Wang C, Yu F, Jin S (2006) Production of ethanol from microwave-assisted alkali pretreated wheat straw. Process Biochem 41:869–873

Chapter 9
Utilization of Paddy Straw for the Production of Hydrolytic Enzymes

Aqsa Tahir, Hafiz Abdullah Shakir, Muhammad Khan, Shaukat Ali, Abdulrahman H. Alessa, Marcelo Franco, and Muhammad Irfan

Abstract *Oryza sativa* commonly called as rice is a second major cereal crop after wheat and is thought to be originated from Asian regions that contribute about 90% of its total global production. Besides feeding populations around globe, it is also producing significant amount of lignocellulosic waste consisting of 28–45% cellulose, 12–32% hemicellulose, and 5–24% lignin along with some other biochemical constituents. Lack of management and awareness among crop producers leads to the open field burning of heaps of paddy straw that is of huge environmental concern, causing air, soil, and water pollution in nearby areas. However, the advent of new biological and technological techniques has led to the efficient in situ and ex situ management of paddy straw. The biotechnological approaches using lignocellulolytic microbes in both solid-state and submerged fermentation conditions to produce important hydrolytic enzymes using paddy straw as a biochemically rich lignocellulosic waste are gaining much attention.

Keyword Hydrolytic enzymes · Paddy straw (PS) · In situ and ex situ management · Lignocellulosic biomass · Fermentation processes · Solid-state fermentation (SSF) · Submerged fermentation (smf) · Microbes

---

A. Tahir · M. Irfan (✉)
Department of Biotechnology, University of Sargodha, Sargodha, Pakistan
e-mail: Irfan.ashraf@uos.edu.pk

H. A. Shakir · M. Khan
Institute of Zoology, University of the Punjab, Lahore, Pakistan

S. Ali
Department of Zoology, Government College University, Lahore, Pakistan

A. H. Alessa
Faculty of Science, Department of Biology, University of Tabuk, Tabuk, Saudi Arabia

M. Franco
Department of Exact Science, State University of Santa Cruz, Ilheus, Brazil

N. Srivastava et al. (eds.), *Paddy Straw Waste for Biorefinery Applications*, Clean Energy Production Technologies, https://doi.org/10.1007/978-981-99-8224-0_9

9.1 Introduction

Rice crop commonly called as paddy is actually a grass (Gramineae) and a member of genus *Oryza*. The rice varieties *Oryza sativa* and *Oryza glaberrima* were originated from Asia and Africa, respectively. Other places like India and Northern Thailand have also been proposed in literature as origins of *O. sativa*. In addition, Yangzi valley of Southern China is also considered among the domesticated places of rice (Dobermann and Fairhurst 2002; Singh et al. 1995). Rice can be grown in a variety of environments. But higher yields are reported in dry season characterized by lesser cloud cover and more photosynthetic active rays in comparison to wet season (Dobermann and Fairhurst 2002). For various geographical areas and environments, the main rice harvesting season lasts from June through October (Zhiqiang et al. 2011). Normally, the growing season of rice crop is almost 260 days, but the most of the modern varieties commonly cultivated took relatively shorter time of about 90–110 days (Dobermann and Fairhurst 2002).

Being the most popular cereal around the globe, rice (*Oryza sativa*) is mostly consumed in the developing countries where it is a major diet component of population (Ramos et al. 2022). For example, in South Asia alone, rice is consumed by around 1670 million individuals (Meetei et al. 2020; Dutta et al. 2020; Urfels et al. 2020). It is estimated by FAO (Food and Agriculture Organization of UN) that 164 Mha of global land is covered by around 760 Mt of rice that are produced each year (FAO 2022). According to regional distribution, about 90% of the total global rice is produced by Asia, and the remaining is contributed by Africa and Latin America. In addition some quantity of this cereal crop is also produced in Europe but only in Mediterranean regions (Ramos et al. 2022). Specifically, 2/3 of the European rice consumption is supported mainly by European Union and Spain where 7.83 million tons of rice crop is being grown per annum (AEE_2022_WEB n.d.).

Considering abovementioned figures along with the world's rice consumption, the post-harvest rice processing produces a blend of lignocellulosic wastes such as paddy straw (PS) in the fields and rice husk during milling process (Ramos et al. 2022). As a most consumed cereal after wheat, rice generates significant amount of waste every year (972 t/anum). According to estimates, during each harvesting season, around 6 t/ha of paddy straw (PS) is produced that is of huge environmental concern (Torregrosa et al. 2021). The production of paddy straw (PS) annually ranges from 370 to 520 million tons, making it a prevalent agri-waste globally (Van Hung et al. 2020). One of the major causes of this enormous waste production is the usage of combined harvesters in the rice fields that produce more waste as compared to manual harvesting. This is because of their inability to cut rice stems deeply, hence resulting in loose straw that offer great difficulties in operation of agricultural machinery for sowing of subsequent crop (Nagar et al. 2020).

In rural populations, rice cultivation not only provides employment opportunities but also ensures the food availability (Bhaduri et al. 2017; Nandan et al. 2021; Taneja et al. 2019). But because of the lack of adequate awareness and poor management of paddy straw (PS) generated in the fields, the farmers are left with

Table 9.1 List of air and soil pollutants produced by open-field burning of paddy straw (modified from Bressan et al. 2022)

	Pollutants			
	Air pollutants		Soil pollutants	
Sr. No.	Names	kg/ha	Names	kg/ha
1	CO	212	Al_2O_3	13.8
2	NO_x	18.9	CaO	4.5
3	N_2O	0.4	Fe_2O_3	3
4	SO_2	12.2	K_2O	13
5	CH_4	7.3	MgO	92
6	NMHC	24.4	MnO	0.9
7	PAH	0.11	Na_2O	6.1
8	Total PM	79.2	P_2O_5	45.7
9	PM 10	22.6	SiO_2	973.2
10	PM 2.5	78.9		

PM particulate matter, *NMHC* non-methane hydrocarbons, *PAH* polycyclic aromatic hydrocarbons

the only option of burning this waste in the open fields. According to literature the crop residues that are burnt every year majorly constitutes waste streams from rice (40%) and then wheat (22%) and sugarcane (20%) (Nagar et al. 2020).

Even in this era of agricultural innovations, environmental sustainability is at greater risk due to the burning of crop residues (Bimbraw 2019; Mondal et al. 2020). High fiber content of paddy straw makes it less susceptible to biodegradation. So in order to get rid of heaps of paddy straw, burning has been considered as a cheap, affordable, and easy traditional solution but with greater social and environmental effects. For instance, open-field burning of paddy straw (PS) results in smoke that upon interaction with weather conditions cause certain respiratory diseases in nearby population. In addition to smoke, this incomplete combustion also results in emission of highly volatile and polluting organic compounds. Besides, it causes emission of several atmospheric particles. These particles badly affect the local hydrobiological cycle, climate, and aerial clarity (Grillo et al. 2020; Singh et al. 2021; Singh et al. 2020b). The pollutants that badly affect air and soil are also mentioned in Table 9.1. It is believed by farmers that PS burning is advantageous as it is helpful in the preparation of fields for subsequent cropping. It is reported by researchers that burning of PS helps in elimination of phytopathogens present in the soil (Chen et al. 2019) with reduced energy costs and results in efficient residue removal in less time (Trivedi et al. 2017; Singh et al. 2020a). On the other hand, the most common disadvantages of PS burning include air pollution and loss of microbial diversity in soil which also adversely affects agricultural sustainability (Goncharov et al. 2020). In addition, it eliminates crucial agricultural resource, namely, straw. Furthermore, it cause degradation of topsoil that lead to reduced arability of landscapes which can be replenished by efficiently recycled paddy straw (Raheem et al. 2019). Advances in science helps to avoid all these problems by both in situ and ex situ management of PS as presented in Fig. 9.1 and production of various valuable products from it (Singh and Brar 2021). Besides these, it can also be utilized in fermentation

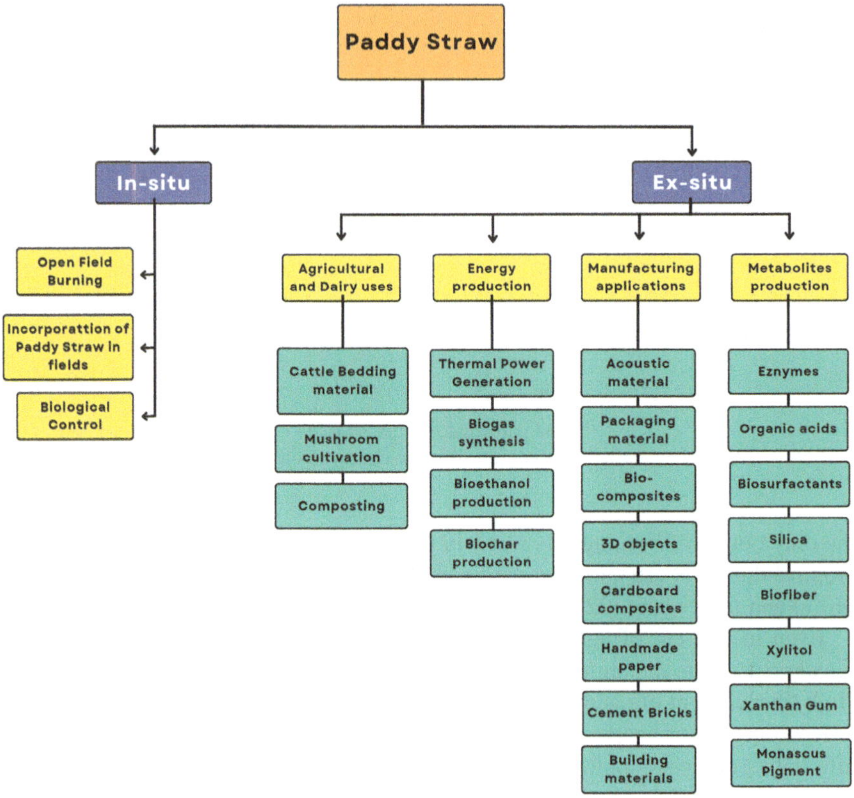

Fig. 9.1 In situ and ex situ paddy straw management methods

processes for the production of various multi-functional hydrolytic enzymes like cellulase, xylanase, laccase, endoglucanase, β-glucosidase, etc. The enhanced production of these enzymes is reported in the presence of medias supplemented with other residues and co-cultures of different microbes (Dhillon et al. 2011; Iyer and Chattoo 2003). The main focus of this chapter is on the production of hydrolytic enzymes utilizing paddy straw (PS) as a substrate in fermentation processes.

9.2 Paddy Straw as a Substrate

Morphologically, PS consist of cylindrical shaped stalk/stem having length ranging from 60 to 120 cm with flat-shaped and elongated leaves that are distributed alternately along the stem. Structural analysis of plant fibers shows that its complex matrix is buildup of hemicellulose, cellulose, and lignin along with other biologically active compounds. Cellulose and hemicellulose bound together by hydrogen bonds provide support to lignin which in turn serve as a natural glue. Hence, the individual plant cells in the form of this structural network provide great strength to plant (Ramos et al. 2022).

Paddy straw is actually a lignocellulosic biomass consisting of three polycarbohydrates among which hemicellulose and cellulose are aliphatic, while lignin is a natural polycyclic aromatic biopolymer. These hydrocarbon polymers can easily be distinguished on the basis of their constituent sugars. For instance, cellulose is a long linear biopolymer of β-glucose subunits that are linked by one to four glycosidic linkages. It is found both in nature and in agro-industrial wastes. Crystalline fibers are formed by high molecular mass cellulose (Razali et al. 2022). On the basis of structural configuration of carbon atoms and hydrogen bonds, cellulose is categorized as primary (trees), secondary (plants), tertiary (agri-food wastage), and quaternary (algae, bacteria, and certain marine organisms). In plants, cellulose being trapped between lignin and hemicellulose is quite difficult to separate. In addition, cellulose is abundant, cheap, and non-toxic, and owing to its complex structure, it is insoluble in some common solvents like water (Razali et al. 2022).

On the other hand, hemicellulose is a shorter heteropolymer consisting of several different polysaccharides including pentoses and hexoses like galactose, arabinose, mannose, xylose, and rhamnose and uronic acids (Goodman 2020; Qaseem et al. 2021). The structure of hemicellulose is relatively complex. These polymers have relatively low molecular weight as compared to cellulose which make them susceptible to hydrolysis under mild conditions. These are second most abundant compound that are found commonly in vegetable fibers. Owing to OH group in the structure, it is hygroscopic in nature (can form bonds with H_2O). In addition, it has less degree of polymerization (80–200) than cellulose (Qaseem et al. 2021). Furthermore, according to variation in structural configuration, hemicellulose are divided into four groups, namely, xyloglucans, mannoglycans, xyloglycans, and β-glucans, with mixed linkages showing diversity in ramifications and different chains and bonds (Huang et al. 2021).

In contrast, the third biopolymer is lignin which is an aromatic hydrocarbon polymer that is composed of oxidativley coupled 4-hydroxyphenylpropanoids usually p-coumaric with synapyl and coniferyl alcohols (Ralph et al. 2004). These abovementioned polymers form very complex and stable 3D structure as illustrated in Fig. 9.2. This structure is commonly named as lignocellulose in the plant cell wall (Goodman 2020). Paddy straw also contains some amount of polyphenols. These are found naturally as secondary metabolite of plants and contain at least a single aromatic ring bound to one or more OH groups. These compounds prevent cellular oxidation and play a role in defense mechanism of plants. According to their structure, polyphenols are categorized mainly into flavonoids, stilbenes, and phenolic acids, but different others can also be found in PS (Khosravi and Razavi 2020).

In addition, minor fractions of some other soluble and insoluble components like proteins, pectin, waxes, and minerals are also present. Certain parameters like crop cultivars and varieties, soil quality, growth stage, environmental conditions, and other factors cause fluctuations in the proportion of these constituents (Kumar et al. 2018). This biochemically rich nature of paddy straw makes it a suitable substrate for

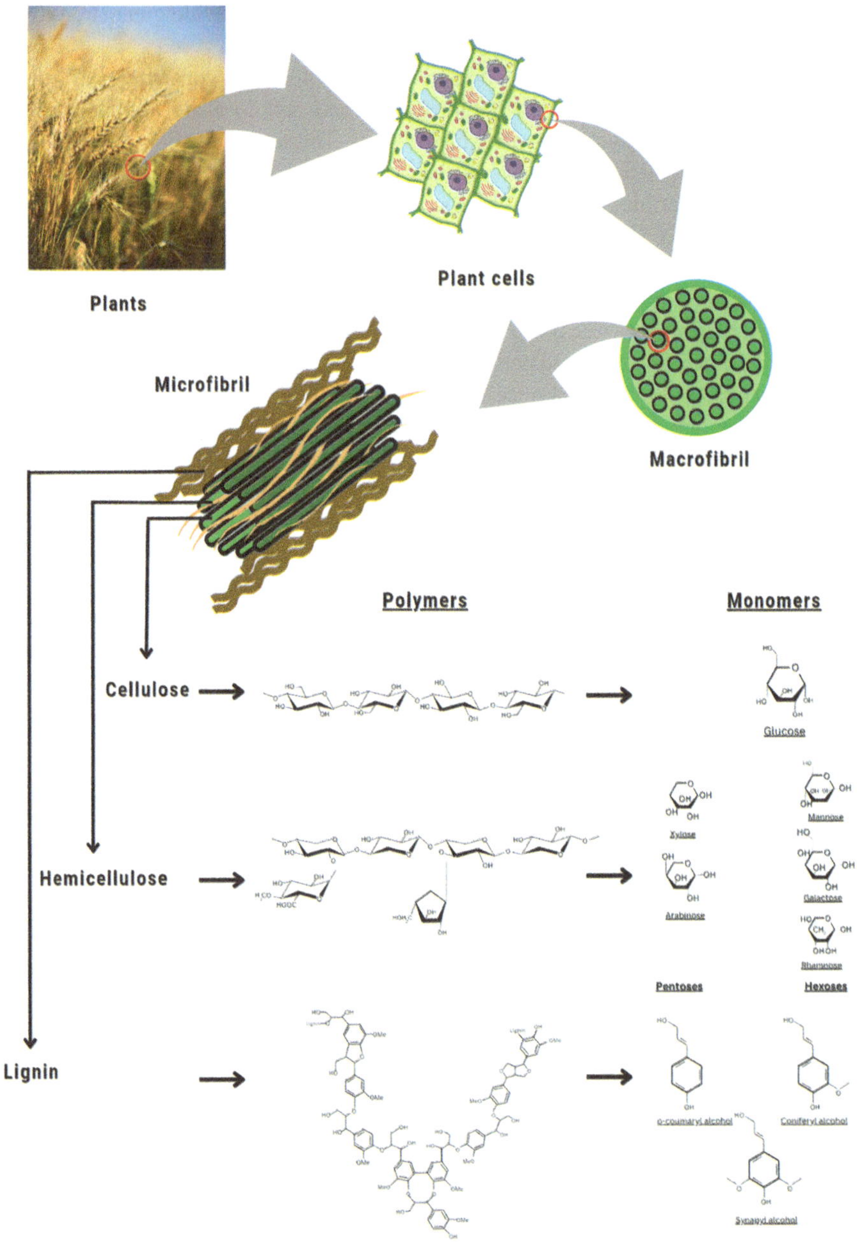

Fig. 9.2 Structure illustration of lignocellulosic biomass along with constituting biopolymers and their monomers

Table 9.2 Biochemical % composition of paddy straw (PS)

Components	% Composition (%)	References
α cellulose	28–45	Kaur et al. (2017)
Hemicellulose	12–32	
Lignin	5–24	
Pentosans	23–28	
Moisture	3.30–10.97	Kumar et al. (2021)
TS	89.03–96.70	
VS in TS	73.28–95.26	
Ash content	4.78–26.72	
Carbon	31.00–47.00	
Hydrogen	4.61–5.40	
Nitrogen	0.28–1.39	
Sulfur	0.14–0.72	
Oxygen	50.46–59.98	

TS total solids, *VS* volatile solids

enzyme production. The complete percentage compositions of all the components present in PS are depicted in Table 9.2.

For its use in fermentation processes, PS must first be converted into simpler forms through several pretreatment methods, because its complex components cannot be utilized efficiently by enzyme producing microbes. The stable complex structure of lignocellulose and other components makes it difficult to separate the individual subunits from the polymers present in plant cell wall, although various chemical, physical, or biological treatments are used individually or in combination as displayed in Fig. 9.3 to convert polymers into simpler reducing sugars and phenolic subunits. However, efficient utilization of PS-derived biomass is yet under development (Wi et al. 2013). So accumulation of PS in soil and open field burning are still commonly practiced despite their hazards to environmental sustainability (Goodman 2020).

9.3 Microbes Utilized in Process

Different types of microbes can be used in fermentation process including bacteria, fungi, yeast, and some others. Due to lignocellulosic nature of paddy straw, those microbes are preferentially utilized in fermentation process that have ability to degrade rigid complex structure of lignocellulose and to use its reduced subunits as a carbon source to produce various desired valuable bioproducts. This chapter mainly focus on the production of hydrolytic enzymes from fermentation of paddy straw. The microbes that can be used for this purpose are discussed in this section.

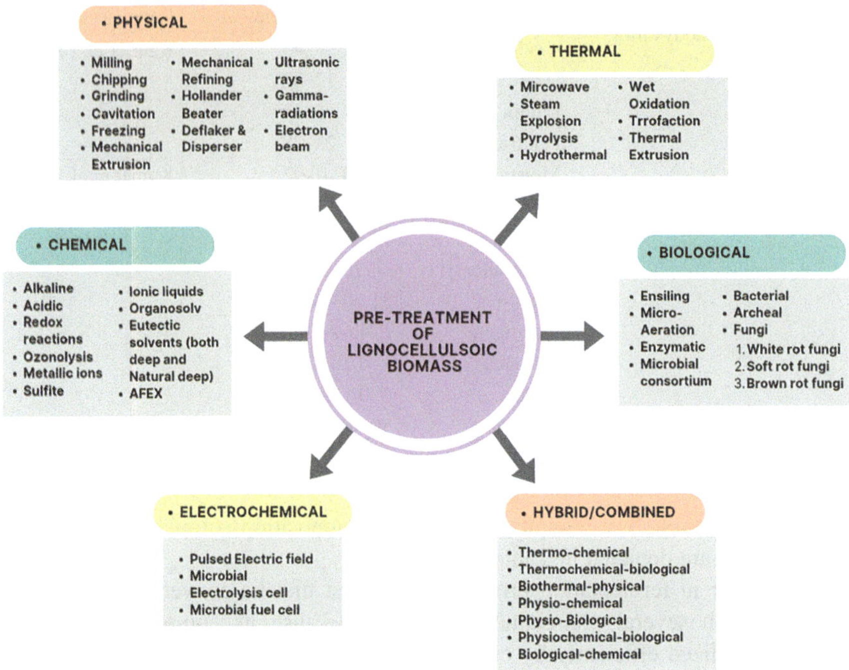

Fig. 9.3 Different types of pretreatment methods used for lignocellulosic biomass

9.3.1 Bacteria

Bacteria being the earliest and most basic form of life on earth play a crucial role in cycling nutrients and maintaining earth's ecology (Ho et al. 2020). It also plays significant role as a producer microbe in fermentation processes. And due to the accessibility and economy of lignocellulosic biomass, it is commonly used in fermentation media as carbon source. So in order to break down lignocellulose, bacteria use its cellulases and hemi-cellulases (Taha et al. 2015a, 2015b). Bacteria are effective microbe for this purpose because of several advantages like ease of culturing, possibilities of enhanced expression and accelerated production (Taha et al. 2015a, 2015b), shorter generation times (Muaaz-Us-Salam et al. 2020), etc. In addition, their metabolic versatility makes them able to endure environmental stress, e.g., fluctuations in pH, temperature, oxygen availability, and salinity (Daniel and Nilsson 1998). Also a recently discovered advantage is that during the latter stages of lignocellulose breakdown, bacteria might show accelerated growth which is beneficial because these stages are renowned for having components that are difficult to hydrolyze. These accelerated growth rates leads to enhanced enzyme production. Bacterial lignocellulases work in the form of multi-enzymatic complexes that are affective for complicated degradation of complex biomass (López-Mondéjar et al. 2019).

Although bacteria use a variety of methods to degrade lignocellulosic biomass, free-enzyme system is most commonly used among them. Aerobic bacteria largely use this system of free enzymes. On the other hand, anaerobic bacteria mostly utilize intricate protein complexes such as xylanosomes and cellulosomes which play role as supportive enzymes to hydrolyze complex biomass (Malgas et al. 2017).

As cellulose is most abundant and common biomolecule, so the microbes that degrade it are actually playing significant role in carbon flux in ecosphere (McDonald et al. 2012). The cellulose degrading bacteria (cellulotyic bacteria) that are mostly isolated belongs to two phylums *Actinobacteria* (order *Actinomycetales*) that are aerobic and *Firmicutes* (order *Clostridiales*) that are anaerobic. Among these two, *Actinobacteria* efficiently degrade cellulose because the other one is incapable of penetrating cellulosic substances (Chukwuma et al. 2021). Mechanism of cellulose degradation in aerobic and anaerobic bacteria is slightly different (Mohee et al. 2008). Aerobic bacteria do so by the action of free-cellulolytic enzymes in two steps (Singhvi and Gokhale 2019). In first step (depolymerization), it converts cellulose present in biomass into cellobiose; then in second step (fermentation), it hydrolyzes cellobiose into organic acids, hydrogen, and carbon dioxide (Hassan et al. 2019). After this, in the latter stages, bacteria dominating in the medium produce valuable products by utilizing these secondary products as carbon and energy source (Beaton et al. 2019). On the other hand, anaerobic cellulose fermentation also takes place in two steps. It involves conversion of sugars into acids or alcohol in first step and then production of biogas from these acids and/or alcohols (Hassan et al. 2019).

Following cellulose, the second most common macromolecule is lignin (Liao et al. 2020) which cannot be easily hydrolyzed and is a major deterrent to lignocellulose degradation. *Streptomyces* has been commonly identified as ligninolytic bacteria that belong to *Actinobacteria*. Some other types of bacteria that can breakdown both lignin and carbohydrate content in lignocellulose include *Thermobifida fusca*, *Caldicellulosiruptor bescii*, and *Clostridium thermocellum* (Lee et al. 2019). Lignolytic bacteria breakdown lignin by three ways: cavitation, tunneling, and erosion (Berg and Laskowski 2005). But the bacteria such as *Proteobacteria* and *Actinobacteria* do lignin degradation by depolymerization, catabolism of aromatic compounds, and biosynthesis of specific product. Depolymerization of lignin is distinct from hemicellulose and cellulose because of the involvement of electron transfer and redox reactions in it (Xie et al. 2016). The bacterial strains used commonly for paddy straw bioconversion are mentioned in Table 9.3.

9.3.2 *Fungi*

Fungal species that can degrade lignin include mostly filamentous fungi. These can be isolated from soil, plants, and lignocellulosic wastes where they are found indigenously. It is reported in various studies that *brown-* and *white-rot fungi* display effective lignocellulolytic capabilities. They have been found to degrade various lignocellulosic biomass like softwood, wheat straw, wood chips, and Bermuda grass

Table 9.3 Strains of bacteria that can convert paddy straw into various hydrolytic enzymes

Strain name	% of degradation	Fermentation state	Time	Enzymes	References
Firmicutes, *Proteobacteria*	–	Submerged fermentation	84 h	Cellulase and xylanase	Taha et al. (2015a)
Aeromonas hydrophila, *Streptomyces thermoviolaceus*, *Pseudomonas poae*, *Bacillus amyloliquefaciens*, *Klebsiella oxytoca*	100	Submerged conditions	72 h	Strawase, cellulase	Taha et al. (2015b)
Bacillus, *Streptococcus*, *Enterococcus*, *Sediminibacterium*, *Lactococcus*, *Rhodanobacter*, *Afipia*, *Ralstonia*, *Alkaliphilus*, *Burkholderia*, *Geobacillus*, and *Erwinia*	–	Submerged state	4 days	Cellulase	Hu et al. (2017)
Streptomyces sp. *MDS*	6	Solid-state conditions	6 days	Endoglucanase, exoglucanase, cellobiases, filter pa-perase, amylase, and xylanase	Saratale et al. (2017)
Paenibacillus polymyxa ND25	–	Submerged fermentation	48 h	Endoglucanase, exoglucanase and β-glucosidase	Bohra et al. (2018)
Sphingobacterium sp. *ksn-11*	60	Submerged fermentation	24 h	Cellulase, xylanase, pectinase, mannanase, and laccase	Neelkant et al. (2019)
Lactobacillus plantarum RI 11	–	Both solid-state and submerged conditions	7 days	Endoglucanase, exoglucanase, β-glucosidase, and mannanase	Zabidi et al. (2020)
Alcaligenes, *Parabacteroides* *Clostridium*, *Lysinibacillus*, and *Sphingobacterium*	71	Submerged conditions	20 days	Endo-glucanase	Zheng et al. (2020)

Table 9.4 Fungal species that are reported to produce hydrolytic enzymes utilizing paddy straw

Fungal species	State of fermentation	Incubation time	Enzyme	References
Trichoderma reesei and *Aspergillus awamori*	Submerged fermentation	5 days	Cellulase	Naher et al. (2021)
Streptomyces psammoticus	Solid-state fermentation	48 h	Laccase	Niladevi et al. (2007)
Aspergillus niger ITBCC L74	Solid-state fermentation	4 days	Cellulase	Maftukhah and Abdullah (2018)
Trichoderma reesei and *Humicola insolens*	Submerged fermentation	7 days	Cellulase, xylanase, and beta-glucosidase	Kogo et al. (2017)
Ganoderma lucidum (white-rot fungi)	Submerged fermentation	5 days	Laccase	Yuliana et al. (2020)
Fomitopsis meliae CFA 2 (brown-rot fungi)	Solid-state fermentation	225.17 h	Endoglucanase	Patela et al. (2021)
Aspergillus fumigatus NITDGPKA3	Submerged fermentation	5 days	Xylanase and cellulase	Sarkar and Aikat (2014)
Aspergillus flavus	Solid-state fermentation	7 days	Protease	Muthulakshmi et al. (2011)

(Alexandropoulou et al. 2017; Cohen et al. 2017; Mishra et al. 2017). In ecosystem, fungal species are the main degraders of wood in the forests. Also various fungal species are used especially for their greater lignolytic efficiencies. For example, *white-rot fungus* can breakdown lignin, hemicellulose, and cellulose. On the other hand, *brown-rot fungus* degrades hemicellulose and cellulose polymers while leaving the tougher lignin walls intact (Tsegaye et al. 2019). Despite their greater lignin degrading potential, fungi are not suitable for larger-scale operations because of their requirement for longer residence time which leads to increasing demands of larger space and higher production costs. However, chemical-free operation, eco-friendliness, and ability to work at mild temperature make them economic alternative producer microbe in fermentation processes (Tsegaye et al. 2019).

In fungi, the mechanism of degradation of lignocellulosic biomass is classified into two types, namely, oxidative and hydrolytic degradation. During oxidative-type lignin degradation, reactive species of oxygen in the form of free radicals such as hydroxyl are produced (Hammel et al. 2002). First, the mutual catalytic action of three enzymes including glyoxaline oxidase, pyranose-2 oxidase, and aryl-alcohol oxidase produces hydrogen peroxide (Martínez et al. 2009). Then these hydrogen peroxides react with iron to produce hydroxyl radicals (Fenton reaction) which act on lignin to degrade it into products of low molecular mass (Hammel et al. 2002). Some other enzymes like laccases and manganese peroxidase also catalyze lignin degradation in oxidative-type reactions (Eggert et al. 1997). In contrast, the mechanism of hydrolytic type of reaction involves the breakdown of glycosidic linkages by means of hydrolytic enzymes (Feijoo et al. 2008). Fungal species used for production of enzymes from paddy straw are indicated in Table 9.4.

9.3.3 Yeast

Saccharomyces cerevisiae commonly called as yeast is mostly used in fermentation processes. However, its inability to utilize hexose and lower metabolic activity at higher concentration of ethanol make it less suitable for fermentation operations involving lignocellulosic biomass. This is a major hindrance in its selection as producer microbe in fermentation processes. Larger-scale production of enzymes using paddy straw as lignocellulosic biomass is still poorly developed due to the unavailability of suitable producer microbe that can efficiently utilize reducing sugars like hexose and pentose (Gabriel and El-halwagi 2013).

9.4 Fermentation Process

The metabolic process of fermentation is used to convert carbohydrates to various valuable metabolites. Paddy straw is lignocellulosic in nature and is used by microbes as a carbon source in fermentation process to produce various hydrolytic enzymes as demonstrated in Fig. 9.4. On the basis of physical state of medium, fermentation is categorized into two types: solid-state and submerged fermentation. The characteristics of these types of fermentation and the microbes utilized in them are discussed below.

9.4.1 Submerged Fermentation (SmF)

SmF is a biochemical process in which microbes are cultivated in liquid medium called broth. Several industrially important enzymes have been produced by this method. This involves first the selection of a suitable microbe and then the preparation of fermentation medium containing sterilized raw material and some nutrients that are essential for the growth of microbes in the presence of sufficient oxygen. Finally, the microbe utilize these nutrients and ultimately degrade them into final

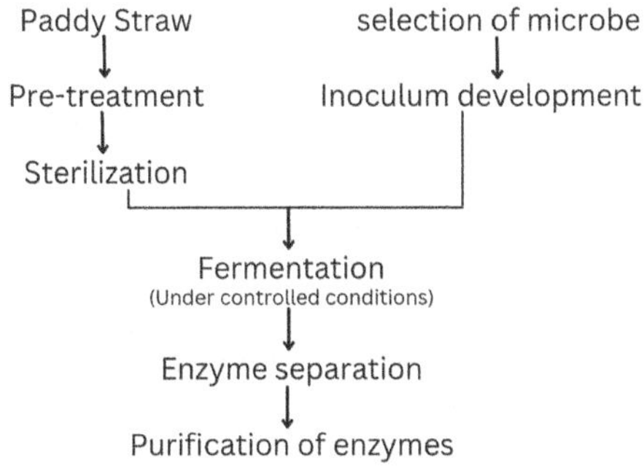

Fig. 9.4 Steps involved in paddy straw fermentation

product usually enzymes that are secreted into the fermentation broth (Renge et al. 2012).

Due to advancements in biotechnology, enzymes are now produced in significant proportions in larger-sized bioreactors that can accommodate volumes of around 1000 cm^3. According to the time limit and operation method, submerged fermentation is further classified into batch, fed-batch, and continuous modes. Batch mode involves the addition of nutrients at the start of reaction only. Fed batch involves the addition of nutrients during the reaction, and continuous mode involves the continuous addition of nutrients and removal of broth from the system at a similar flow rate. One of the advantages of submerged fermentation is that it is a highly controlled process. All the parameters like pH, temperature, oxygen utilization, formation of carbon dioxide, etc. are optimized for efficient working of microbe and enzyme production. In downstream processing, first step mostly involves the removal of insoluble microbial biomass by centrifugation and its recycling and inactivation by its treatment with lime. Most industrially important enzymes are usually extracellular and are secreted in broth from which they can be concentrated by membrane filtration, evaporation, crystallization, etc. according to their desired application. Enzymes can be further purified by techniques like ion or gel chromatography, etc. The final enzyme produced (powdered form) can be used either in solid form (granules) or in liquid form (enzyme solutions) depending upon the required state for particular application. Mostly enzymes are immobilized on the solid support for prolonged usage (Renge et al. 2012).

Various microbes including bacteria and fungi can be used for paddy straw degradation to produce enzymes. Some microbial strains (fungi and bacteria) that can be utilized include species of *Fusarium*, *Aspergillus*, *Phoma*, etc. In addition, *Vibrio*, *Cytophaga*, *Cellulomonas*, *Polyangium*, *Nocardia*, and *Streptomyces* also show cellulolytic activities. Some white-rot fungi like *Phanerochaete chrysosporium* have also been reported in literature for this purpose. These lignolytic and cellulolytic microbes have frequently been utilized for the synthesis of enzymes such as β-glucosidase, exo- and endo-cellulases, peroxidases, laccases, etc. using lignocellulosic wastes like paddy straw in submerged fermentation conditions (Mishra and Pandey Lata 2007).

9.4.2 Solid-State Fermentation (SSF)

The other method that is used for enzyme production is solid-state fermentation. It is a type of fermentation in which microbes are cultivated on a solid support or raw materials that are in solid state with low water content such as paddy straw, wheat bran, sugar bagasse, husks, paper pulp, etc. (Renge et al. 2012; Subramaniyam and Vimala 2012).

Like submerged fermentation, SSF has some advantages like simple operation with less equipment, production of product in high concentration and volumes, and generation of less effluent. These advantages make SSF a suitable alternate to submerged fermentation. In SSF, various substrates can be used, but lignocellulosic substrates like straws, brans, and husks of wheat and rice are preferable. In addition, corn and wheat flour, pulp of sugar beet, etc. are also used. Substrate selection depends upon some factors like availability and cost of substrate. Moisture level and particle size are some other factors. For example, small-sized particles result in better proliferation of microbe on large surface area. However, too small particles cause poor growth, impeded respiration, and hence lower enzyme production as well. Similarly, larger particles result in efficient respiration, but the reduced surface area badly affects other aspects (Renge et al. 2012). In the same way, moisture of substrate also affects the rheology of medium and metabolic activity of microbe in SSF and in turn enzyme production. So, water content should be carefully maintained to minimize its detrimental effects on microbial activity. SSF have widely been employed for the synthesis of vital hydrolytic enzymes like cellulases, pectinases, proteases, and glucoamylases (Renge et al. 2012; Suganthi et al. 2011).

Microbes that are used mostly in SSF include filamentous fungi and yeast. Some species of bacteria are also used. Filamentous fungi that are mostly used include genera of *Fusarium*, *Aspergillus*, *Trichoderma*, *Penicillium*, and *Rhizopus*. Yeast such as *Candida* sp., *Saccharomyces cerevisiae*, and *Saccharomyces boulardii* are employed in SSF. *Actinobacteria* species used for this purpose include *Streptomyces thermonitrificans* and *Streptomyces chattanoogensis* (Hu et al. 2012; Orozco et al. 2008; Munishamanna et al. 2017). Bacterial species that are notably used in SSF include *Bacillus mycoides*, *Bacillus megaterium*, etc. Some species of *Lactobacillus* including *L. plantarum*, *bulgaricus*, *acidophilus*, *delbrueckii*, *coryniformis*, and *rhamnosus* are also reported to be involved in SSF (Oboh 2006; Hongzhang et al. 2011; Hsu et al. 2013; Andriani et al. 2015; Saanu and Oladiti 2018).

9.5 Hydrolytic Enzymes

9.5.1 Cellulases

Cellulase are the group of enzymes that are known for hydrolyzing β-1, 4 glycosidic bonds present in the polymeric cellulose to convert it into glucose molecules. The classes of enzymes that are included in cellulase are beta-glucosidases, endo-glucanases (endo 1, 4-β-d-glucanase), and exo-glucanases (exo 1, 4-β-d-glucanase) (Schülein 1988). Glycosyde hydrolase (GH) is a family of catalytic modules that are further categorized into several sub-classes based upon their 3D structure and sequential arrangement of amino acids. The catalytic modules of cellulase belong to this family. The enzymes of this family hydrolyze glycosidic bonds present in cellulose mainly via mechanism of acid–base catalysis. This catalysis take place by the interaction of two major enzyme residues: a nucelophile and a proton donor present at the regions of active site (Davies and Henrissat 1995). On the basis of difference in spatial arrangement of the catalytic modules in enzyme structure, the mechanism of hydrolysis may be inversion or retention. Three classes of cellulase enzymes containing particular catalytic residues do stepwise hydrolysis of cellulose polymer as follows: (1) endoglucanases contain catalytic residues belonging to GH families of mostly 74, 51, 48, 45, 44, 12, and 5–9. Endoglucanases act on cellulosic polymer to break β-1, 4 linkages to expose its reducing and non-reducing ends. The presence of catalytic module along with carbohydrate-binding module (CBM) is reported mostly in fungal endoglucanases but with few exceptions (Kubicek 2013). On the other hand, bacterial endoglucanases contain multiple catalytic modules with their CBMs (Payne et al. 2015). (2) The next step of hydrolysis is performed by exoglucanases also called as cellobiohydrolases (CBHs) containing catalytic modules that belong to GH families of 74, 48, 9, 7, 6, and 5. These enzymes cause the hydrolysis of polymeric cellulose into cellobiose by acting on the exposed non-reducing or reducing ends in the chain. A great diversity has been observed in both bacterial and fungal CBHs (Naga Padma et al. 2017). (3) The final step of cellulose hydrolysis is performed by β-glucosidases. It catalyzes the cleavage of terminal non-reducing β-d-glucosyl residues as well as the removal of β-d-glucose (Leah et al. 1995). Catalytic modules of 9, 3, and 1 GH families are present in β-glucosidases. Feedback inhibition regulates this cellulolytic process by the interaction of end product glucose with b-glucosidases. The main difference in the structure of CBHs and b-glucosidases is absence of carbohydrate binding module. At the last of this cellulose hydrolysis, glucose units are released from cellobioses (Payne et al. 2015). A diverse range of microorganism, when grown on cellulose-based material, produce cellulases as listed in Table 9.5. Most of the industrial cellulases are produced by microbial sources like fungi and bacteria. Both aerobic bacteria and aerobic fungi show similar cellulose-degrading mechanism. Cellulases are widely used in food, textile, and paper industry. Most cellulases that are used in food industry are obtained from fungal species *Trichoderma* and *Aspergillus*, while bacterial cellulases are obtained from species *Paenibacillus* and *Bacillus* (Sukumaran et al. 2005).

9.5.2 Xylanases

Hemicellulose consists of xylan that can be cleaved by using xylanases. These enzymes are secreted by microorganisms. One of these three enzymes, endoxylanases, exoxylanases, and β-xylosidases, is used to cleave the xylan, component of hemicellulose. The β-1, 4 bonds of xylan core break by endoxylanase (EC 3.2.1.8). Xylooligosaccharides are generated by non-reducing ends of xylan. These xylooligosaccharides are secreted when exoxylanase breaks down the β-1, 4 linkages of xylan. Xylose is secreted when β-xylosidase cleaves the xylooligosaccharides and xylobiose (Sukumaran 2009). Few kinds of xylanases have an extra CBM for interacting to substrates, and the catalytic component performs the majority of the enzyme's important tasks. Carbohydrate esterases (CE) and glycoside hydrolases (GH) are two main catalytic modes of hemicellulose. The xylan core is hydrolyzed by endoxylanase, which includes catalytic cores from

Table 9.5 Cellulases produced by various microbes using paddy straw as a substrate

Microbe	State of fermentation	PH	Temp	Incubation time	References
Aspergillus niger	Submerged fermentation	6.0	20 °C	48 h	Saranraj (2011)
Pleurotus ostreatus (white-rot fungus)	–	4.8	35–45 °C	–	Vijaya and Singaracharya (2005)
Trichoderma reesei	Submerged fermentation	–	–	5 days	Naher et al. (2021)
Aspergillus awamori	Submerged fermentation	–	–	3 days	Naher et al. (2021)
Volvariella volvacea	Submerged fermentation	–	35 ± 2 °C	8 days	Choudhary et al. (2009)
Bacillus subtilis	–	4.0	35 °C	48 h	Anu et al. (2021)
Stenotrophomonas maltophilia	Submerged conditions	7.0	37 °C	72 h	Tamilanban et al. (2017)
Penicillium expansum	Solid-state fermentation	–	30 °C	8 days	Sharifzadeh et al. (2020)
Trichoderma harzianum	Solid-state fermentation	–	32 °C	7–10 days	Karthick Raja Namasivayam et al. (2015)
Aspergillus fumigatus	Solid-state fermentation	4.0	33 °C	90 h	Aikat (2012)

the GH families 8, 10, 11, 30, and 43, with GH 10 and 11 proving to be most prevalent (Collins et al. 2005). The GH10 seems to be more effective on modified xylan, and these have different substrate aspects. They could possibly include CBMs, much like cellulases (Sweeney and Xu 2012). The xylan core is randomly split by exoxylanases from the inner side, generating longer chain xylo-oligomers which then act as substrate for xylosidase enzymes. The mode of action of these enzymes makes them the members of the GH families 3, 30, 39, 43, 52, and 54. Xylanases is the name given to such two enzymes when they are combined. Xylooligosaccharides and xylobiose are affected by xylosidase or xylan-1, 4-xylosidase, which secrete xyloses (Juturu and Wu 2014). Various microbes are capable of producing xylanases as depicted in Table 9.6. Microbes such as actinomycetes, bacteria, and fungus secrete xylanases. *Streptomyces*, *Bacillus*, and *Pseudomonas* are the main actinomycete and bacterial species that produce xylanase (Sanghi et al. 2010; Sharma and Chand 2012). The best temperature for xylanase function is somewhere between 35 and 60 °C, whereas those produced by bacterial and actinobacterial stains are efficient throughout wider pH ranges (5.0–9.0). Because of the elevated concentration and extracellular emission of the enzyme, fungi are important sources of xylanase (Nair et al. 2008). *Aspergillus* species, *Fusarium* species, and *Penicillium* species are the main fungi that secrete xylanase. The fungal xylanases are more catalytically active relative to bacteria or yeast (Mandal 2015).

Table 9.6 List of microbes that can produce xylanases using paddy straw

Microbe	State of fermentation	PH	Temperature	Incubation time	References
Volvariella volvacea	Submerged fermentation	–	35 ± 2 °C	8 days	Choudhary et al. (2009)
Trichoderma harzianum	Solid-state fermentation	–	32 °C	7 to 10 days	Karthick Raja Namasivayam et al. (2015)
Bacillus flexus	Both solid-state and submerged conditions	–	37 °C	48 h	Meurial and Uthandi (2020)
Phanerochaete chrysosporium	Submerged condition	–	–	15 days	Mishra and Pandey Lata (2007)
Cytophaga hutchinsonii	Submerged fermentation	–	–	7 days	Mishra and Pandey Lata (2007)
Bacillus altitudinis	–	7.0	37 °C	72 h	Ketsakhon et al. (2022)
Penicillium oxalicum	Submerged fermentation	–	30 °C	4 days	Zahari et al. (2016)
Promicromonospora sp. *MARS*	Submerged fermentation	8.0	65 °C	48 h	Joshi et al. (2011)
Aspergillus fumigatus	Solid-state fermentation	4.0	33 °C	90 h	Aikat (2012)
Schizophyllum commune	Solid-state fermentation	7.0	30 °C	8 days	Gautam et al. (2018)

9.5.3 Proteases

These are the enzymes that cause the hydrolysis of polypeptides and proteins by breaking the peptide bond present between the amino acid subunits. Proteases are extensively used in various sectors which include food, drug, and detergent industries. In market, about 60% of the enzymes that are used in industries are proteases (Singh et al. 2016). The market demand for this enzyme has been increased globally by 5.3% in the years 2014–2019. And the increasing applications of protease in leather processing and bioremediation might increase its demands further more. Microorganisms (including fungi and bacteria), plants, and animals are considered as major sources from which protease can be obtained. Proteases can be categorized according to their origin, type of reactive groups bound to the catalytic sites, and their catalytic activities. Also according to the site of action along the polypeptide chain, proteases are classified into endopeptidases and exopeptidases (Rao et al. 1998). Endopeptidases hydrolyze the peptide bonds in the inner chain region, while exopeptidases do so at the ends of chain. On the basis of reactive groups linked to the active sites, proteases are categorized into six classes such as metallo, serine, cysteine, threonine, aspartic, and glutamic acid protease (Li et al. 2013). Various proteases that are obtained from plants include ficin, papain, and bromelain that are widely used at industrial scale for food applications including brewing, milk coagulation, and meat tenderization, for dihestive aid, etc. (Patel et al. 2013). Fungal proteases that are acidic in nature have ability to work efficiently in low pH environment hence playing a role in improving beer quality by balancing the profile of its amino acids (Nogent-sur-Seine, France). In the process of cheese manufacturing, proteases are used for the production of macropeptides and para casein by hydrolyzing its peptide bonds at specific sites (Salleh et al. 2006).

9.5.4 α-Amylases

These are hydrolytic enzymes that are capable of degrading starch by the hydrolysis of α-1, 4glycosidic linkages present in polysaccharides. This hydrolysis reaction result in the short-chain product called dextrins (Sindhu et al. 2017). A variety of living organisms are capable of producing α-amylases as indicated in Table 9.7. These are actually metallo-enzymes that require Ca^{+} ions for their structural integrity, stability, and activity (Sindhu et al. 2016). These hydrolytic enzymes are of great industrial importance with variety of applications including brewing, baking, digestive aids, and starch liquefaction (Rodríguez Couto and Ángeles Sanromán 2006). In addition, they play role in manufacturing of branched dextrins of high MW. Powdery foods and rice cakes are also prepared by α-amylases by their glazing action (Aiyer 2005). These enzymes do starch liquefaction as well which involves the conversion of starch polymer into fructose and glucose syrups. This process of starch conversion is comprised of three steps including (1) gelatinization, (2) liquefaction, and (3) saccharification. The granules of starch are dissolved to form a viscous suspension in gelatinization. Then partial hydrolysis of this suspension takes place in liquefaction which results in reduced viscosity. At last, maltose and glucose are produced in the final step of saccharification. This process of saccharification requires α-amylase enzymes of thermostable characteristic that are obtained from *Bacillus stearothermophilus*, *B. licheniformis*, and *B. amyloliquefaciens* (Van der Maarel et al. 2002). In ethanol preparation, α-amylases are involved in the conversion of starch to reduced sugars that are further fermented by *Saccharomyces cerevisiae* to alcohol. The collaborative activities of pectinases and cellulases along with α-amylases contribute to the clarity of fruit juices, yield improvement, and reduce the processing cost as well (Kumar 2015; Garg et al. 2016).

9.5.5 Pectinases

Pectinase are the hydrolytic enzymes that hydrolyze pectic polymers by breaking their glycosidic bonds. Pectin (substrate) is found in the walls/peels of fruits such as tomato, apple, pineapple, lemon, orange, and some other fruits. On the basis of their

Table 9.7 Other important enzymes produced by various microbes utilizing paddy straw substrate

	Microbe	State of fermentation	PH	Temperature	Incubation time	References
Alpha-amylase	*Bacillus subtilis*	Solid-state fermentation	7.0	55 °C	48 h	Hassan and Abd Karim (2012)
	Aspergillus niger	Submerged fermentation	7.0	28 °C	7 days	Kanti and Sudiana (2018)
	Achromobacterx-ylosoxidans	Submerged fermentation	–	37 °C	48 h	Mahalakshmi and Jayalakshmi (2016)
Pectinase	Burkholderia sp SMB1	Submerged conditions	7.0	40 °C	24 h	Beladhadi et al. (2022)
	Bacillus subtilis	Both solid and submerged fermentation	7.0	37 °C	96 h	Kumari et al. (2014)
	Aspergillus niger NCIM 548	Both solid-state and submerged fermentation	4.6 (SmF)	–	126 h (SmF)	Kumar et al. (2011)
			4.8 (SSF)		156 h (SSF)	
Laccase	*Burkholderia* sp *SMB1*	Submerged conditions	7.0	40 °C	24 h	Beladhadi et al. (2022)
	Trichoderma asperellum LBKURCC	Solid-state fermentation	–	30 °C	8 days	Rahayu et al. (2019)
	Streptomyces psammoticus, MTCC 7334	Solid-state fermentation	8.0	32 °C	48 h	Niladevi et al. (2007)
Mananase	*Sphingobacterium* sp. *ksn-11*	Submerged fermentation	7.0	40 °C	–	Neelkant et al. (2019)
	Burkholderia sp. *SMB1*	–	7.0	40 °C	24 h	Beladhadi et al. (2022)

functions, pectinase are classified into pectin esterases (responsible for removal of methoxyl and acetyl groups), polygalacturonases (that catalyze hydrolysis of α-(1–4) glycosidic bonds), pectate lyase, and pectin lyase (Saadoun et al. 2013). These enzymes can be obtained from natural microbes as reported in Table 9.7, but attempts have been made by recombinant technology for their increased yield and thermostability (Rebello et al. 2017). Pectinases can attack both the hairy and smooth regions present in pectin (Pedrolli et al. 2009). On the basis of pH, these are classified as alkaline and acidic pectinases. While on the basis of site of action, these are categorized as endopeptidase (random cleavage of bonds) and exopeptodases (acting on terminal ends). Pectinases have vast range of industrial applications like food processing, bleaching of paper, remediation, etc. (Pasha et al. 2013). Pectinase give juices a relatively clear appearance with filter-ability as compared to other enzyme counterparts (Saadoun et al. 2013).

9.5.6 Laccase

Laccases are diversified group of multi-copper oxidases. They are usually called blue oxidases. These are employed in numerous industries due to their higher capabilities of oxidizing phenolic compounds. These enzymes have many applications as a biocatalyst in synthesizing various chemicals, paper bio-bleaching, bio-sensing, bioremediation, finishing fabrics, and stabilizing wines. They show distinct specificity regarding its substrates and hence have an extensive range of oxidizable substrates which is primarily dependent on the microbial origin of enzymes (Madhavi and Lele 2009). The large number of compounds can be oxidized by laccases which include aromatic-amines, ascorbates, and phenolics (Giardina et al. 2010; Madhavi and Lele 2009). For the purpose of splitting up dioxygen linkage, laccases contain four atoms of copper with four oxidized and four reduced electrons (Giardina et al. 2010). Several microbial species are known to synthesize laccases as mentioned in Table 9.7. Few fungal species are used in production of laccases by secondary metabolism during fermentation (Morozova et al. 2007). *Deuteromycetes*, *Ascomycetes*, and *Basidiomycetes* are well-known for the production of enzyme lacasses (Gochev and Krastanov 2007; Sadhasivam et al. 2008). Laccase can be produced from *Funaliatrogii* a white-rot fungus in absorbent mode of fermentation. *F. rogii* can produce laccase at a maximum yield of 11,900 U/L, which is 4.97 times higher than the output obtained by regular fermentation (Li et al. 2017). Transgenic laccases are produced by *Bacillus licheniformis* for industrial applications (Tonin et al. 2016). In last few decades, laccase production had been achieved by its heterologous expressions in various microbes. For instance, fmb-103 genes of *Bacillus vallismortis* were cloned in BL21 (DE3) cells of *Escherichia coli* to express heterologously (Sun et al. 2017).

9.5.7 *Mannanase*

Mannanases is an enzyme group that breakdown mannan, an integral component found in cell wall of plants along with hemicellulose (Guan et al. 2018). Mannanases is group of three enzymes that play role in hydrolyzing linear mannans. These enzymes include beta-mannanases or 1, 4-β-D mannohydrolases, beta-glucosidases or 1, 4-β-D glucoside glucohydrolases, and beta-mannosidases also called 1,4-β-D mannopyranoside hydrolases (Chauhan et al. 2012). β-Mannanases catalyze the endo-hydrolysis of mannan chains and produce short-chain products such as β-1,4-manno-oligosaccharides also called mannobiose by breaking inner glycosidic bonds (McCleary and Matheson 1983). Further hydrolysis is catalyzed by β-mannosidases. This enzyme, by its exo-hydrolysis activity, attacks the non-reducing ends of mannan polymer, hence hydrolyzing the mannobiose complex (disaccharide) into separate mannose (monomer) units (Gomes et al. 2007). Beta-glucosidase hydrolyzes the oligomers generated from hydrolysis of galactoglucomannan and glucomannan. It attacks the non-reducing terminals of their oligomers and results in individual β1, 4-glucopyranose units (Mamma et al. 2004). Several species of bacteria and fungi are reported in literature for production of variety of mannanases. Among *Bacillus* spp., different strains of *B. subtilis* produce mannanases. Some fungal organisms like *Aspergillus* spp. are reported as producers of mannan-degrading enzymes (Dhawan and Kaur 2007). Some other fungal and bacterial species that are known as mannanase producers include *Streptomyces* spp., *Penicillium* spp., and *Clostridium* spp. (Chauhan et al. 2012). Mannanase is attracting great attention in pulp and paper industries owing its hemicellulolytic capability (Clarke et al. 2000). In addition, mannanases also have a multitude of applications in textile, feed, oil, and food industries (Christgau et al. 1994; Naganagouda et al. 2009). The synthesis of mannanase utilizing lignocellulosic feedstock has also been reported by several researchers and is shown in Table 9.7.

9.6 Conclusion

The biochemically rich nature of paddy straw makes it a potential substrate for its bio-processing into numerous important biological products like hydrolytic enzymes that are significant part of most of the industrial processes. Environmental pollution and inefficient management strategies stimulate the utilization of PS in fermentation processes. Isolation of new strains of microbes having high lignocellulolytic potential and advancement in pretreatment methods along with improvements in fermentation technology has paved the way for efficient utilization of paddy waste. Although various tests have been proven effective for production of hydrolytic enzymes from PS, most of these finding are confined to lab-scale level and need further developments in order to be scaled up. In order to scale up these processes to industrial level, interdisciplinary efforts involving microbiology, biotechnology, mechanical engineering, software engineering, and analytical chemistry may be required keeping in view the economics of whole process of manufacturing and the end product cost in market.

References

AEE_2022_WEB (n.d.) Volume of grain paddy rice produced in the different Spanish autonomous communities in 2020. https://es.statista.com/estadisticas/501804/produccion-de-arroz-cascara-grano-en-comunidades-autonomas-de-espana/. Accessed 26 Aug 2023

Aikat (2012) Cellulase and xylanase production from rice straw by a locally isolated fungus Aspergillusfumigatus NITDGPKA3 under solid state fermentation—statistical optimization by response surface methodology. J Technol Innov Renew Energy. https://doi.org/10.6000/1929-6002.2012.01.01.7

Aiyer PV (2005) Amylases and their applications. Afr J Biotechnol 4(13):1525–1529

Alexandropoulou M, Antonopoulou G, Fragkou E, Ntaikou I, Lyberatos G (2017) Fungal pretreatment of willow sawdust and its combination with alkaline treatment for enhancing biogas production. J Environ Manag 203:704–713. https://doi.org/10.1016/j.jenvman.2016.04.006

Andriani Y, Safitri R, Abun (2015) Improvement protein quality of cassava peels by solid-state fermentation using cellulolytic microbial consortium. Sci Pap Anim Sci Ser 63:250–253

Anu, Kumar S, Kumar A, Kumar V, Singh B (2021) Optimization of cellulase production by *Bacillus subtilis* subsp. *subtilis* JJBS300 and biocatalytic potential in saccharification of alkaline-pretreated rice straw. Prep Biochem Biotechnol 51:697–704. https://doi.org/10.1080/10826068.2020.1852419

Beaton D, Pelletier P, Goulet RR (2019) Microbial degradation of cellulosic material and gas generation: implications for the management of low-and intermediate-level radioactive waste. Front Microbiol 10:204

Beladhadi R, Shankar K, Jayalakshmi S, Sreeramulu K (2022) Production of cocktail of lignolytic, cellulolytic and hemicellulolytic enzymes by the novel bacterium burkholderiasp SMB1 utilizing rice bran and straw: application in the saccharification of untreated agro-wastes for bioethanol production. Waste Biomass Valor 13(3):1–13. https://doi.org/10.1007/s12649-021-01607-7

Berg B, Laskowski R (2005) Decomposers: soil microorganisms and animals. Adv Ecol Res 38:73–100

Bhaduri D, Purakayastha TJ, Patra AK, Singh M, Wilson BR (2017) Biological indicators of soil quality in a long-term rice-wheat system on the Indo-Gangetic Plain: combined effect of tillage-water-nutrient management. Environ Earth Sci 76:202

Bimbraw AS (2019) Generation and impact of crop residue and its management. Curr Agric Res J 7: 304–309

Bohra V, Dafale NA, Purohit HJ (2018) Paenibacilluspolymyxa ND25: candidate genome for lignocellulosic biomass utilization. 3 Biotech 8:248

Bressan M, Campagnoli E, Ferro CG, Giaretto V (2022) Rice straw: a waste with a remarkable green energy potential. Energies 15:1355. https://doi.org/10.3390/en15041355

Chauhan PS, Puri N, Sharma P, Gupta N (2012) Mannanases: microbial sources, production, properties and potential biotechnological applications. Appl Microbiol Biotechnol 93:1817–1830

Chen JH, Gong YZ, Wang SQ, Guan BZ, Balkovic J, Kraxner F (2019) To burn or retain crop residues on croplands? An integrated analysis of crop residue management in China. Sci Total Environ 662:141–150

Choudhary M, Dhanda S, Kapoor S, Soni G (2009) Lignocellulolytic enzyme activities and substrate degradation by volvariella volvacea, the paddy straw mushroom/Chinese mushroom. Indian J Agric Res 43(3):223–226

Christgau S, Andersen L, Kauppinen S, Heldt-Hansen H, Dalboege H (1994) Enzyme exhibiting mannanase activity. Patent WO1994025576A1, 10 Nov 1994

Chukwuma OB, Rafatullah M, Tajarudin HA, Ismail N (2021) A review on bacterial contribution to lignocellulose breakdown into useful bio-products. Int J Environ Res Public Health 18:6001. https://doi.org/10.3390/ijerph18116001

Clarke JH, Davidson K, Rixon JE, Halstead JR, Fransen MP, Gilbert HJ, Hazlewood GP (2000) A comparison of enzyme-aided bleaching of softwood paper pulp using combinations of xylanase, mannanase and alpha-galactosidase. Appl Microbiol Biotechnol 53:661–667

Cohen M, Lepesant G, Lamari F, Bilodeau C, Benyei P, Espadas-tormo I, Jaboeuf D, Alonso-rold M, Belandria V, Silar P, Dicko M (2017) Biomolecules from olive pruning waste in Sierra Magina engaging the energy transition by multi-actor and multidisciplinary analyses. J Environ Manag 216:204–213. https://doi.org/10.1016/j.jenvman.2017.03.067

Collins T, Gerday C, Feller G (2005) Xylanases, xylanase families and extremophilicxylanases. FEMS Microbiol Rev 29(1):3–23. https://doi.org/10.1016/j.femsre.2004.06.005

Daniel G, Nilsson T (1998) Developments in the study of soft rot and bacterial decay. In: Bruce A, Palfreyman JW (eds) Forest products biotechnology. CRC Press, Boca Raton, FL, pp 37–62

Davies G, Henrissat B (1995) Structures and mechanisms of glycosyl hydrolases. Structure 3:853–859. https://doi.org/10.1016/S0969-2126(01)00220-9

Dhawan S, Kaur J (2007) Microbial mannanases: an overview of production and applications. Crit Rev Biotechnol 27:197–216

Dhillon GS, Oberoi HS, Kaur S, Bansal S, Brar SK (2011) Value-addition of agricultural wastes for augmented cellulase and xylanase production through solid-state tray fermentation employing mixed-culture of fungi. Ind Crop Prod 34:1160–1167

Dobermann A, Fairhurst T (2002) Rice straw management. Better Crops Int 16:1–11

Dutta SK, Laing AM, Kumar S, Gathala MK, Singh AK, Gaydon DS, Poulton P (2020) Improved water management practices improve cropping system profitability and smallholder farmers' incomes. Agric Water Manage 242:106411

Eggert C, Temp U, Eriksson KL (1997) Laccase is essential for lignin degradation by the white rot fungus pycnoporuscinnabarinus. FEBS Lett 407:89–92. https://doi.org/10.1016/S0014-5793(97)00301-3

FAO (2022) Cultivos y productos de ganadería. https://www.fao.org/faostat/es/#data/QCL

Feijoo G, Moreira MT, Alvarez P, Lú-Chau TA, Lema JM (2008) Evaluation of the enzyme manganese peroxidase in an industrial sequence for the lignin oxidation and bleaching of eucalyptus Kraft pulp. J Appl Polym Sci 109(2):1319–1327. https://doi.org/10.1002/app.28084

Gabriel KJ, El-halwagi MM (2013) Modeling and optimization of a bioethanol production facility. Clean Tech Environ Policy 15(6):931–944. https://doi.org/10.1007/s10098-013-0584-8

Garg G, Singh A, Kaur A, Singh R, Kaur J, Mahajan R (2016) Microbial pectinases: an ecofriendly tool of nature for industries. 3 Biotech 6:47. https://doi.org/10.1007/s13205-016-0371-4

Gautam A, Kumar A, Bharti A, Dutt D (2018) Rice straw fermentation by Schizophyllum commune ARC-11 to produce high level of xylanase for its application in pre-bleaching. J Genet Eng Biotechnol 16:2. https://doi.org/10.1016/j.jgeb.2018.02.006

Giardina P, Faraco V, Pezzella C, Piscitelli A, Vanhulle S, Sannia G (2010) Laccases: a never-ending story. Cell Mol Life Sci 67:369–385. https://doi.org/10.1007/s00018-009-0169-1

Gochev VK, Krastanov AI (2007) Isolation of laccase producing Trichoderma sp. Bulg J Agric Sci 13:171–176

Gomes J, Terler K, Kratzer R, Kainz E, Steiner W (2007) Production of thermostable β-mannosidase by a strain of thermoascusaurantiacus: isolation, partial purification and characterization of the enzyme. Enzym Microb Technol 40:969–975

Goncharov AA, Gongalsky KB, Yazrikova TE, Kostina NV, Korobushkin DI, Makarov MI, Zaitsev AS (2020) Greenhouse gas-producing soil biological activity in burned and unburned forests along a transect in European Russia. Appl Soil Ecol 148:103491

Goodman BA (2020) Utilization of waste straw and husks from rice production: a review. J Bioresour Bioprod 5:143–162. https://doi.org/10.1016/j.jobab.2020.07.001

Grillo G, Tabasso S, Cravotto G, Van Ree T (2020) Burning biomass: environmental impact on the soil. In: Mammino L (ed) Biomass burning in Sub-Saharan Africa. Springer, Dordrecht, pp 15–30

Guan W, Xu G, Duan J, Shi S (2018) Acetone–butanol–ethanol production from fermentation of hot-water-extracted hemicellulose hydrolysate of pulping woods. Ind Eng Chem Res 57:775–783

Hammel KE, Kapich AN, Jensen KA Jr, Ryan ZC (2002) Reactive oxygen species as agents of wood decay by fungi. Enzym Microb Technol 30:445–453

Hassan H, Abd Karim K (2012) Utilization of agricultural by-products for alpha-amylase production under solid state fermentation by bacillus subtilis. Eng J 16(5):177–185. https://doi.org/10.4186/ej.2012.16.5.177

Hassan SS, Williams GA, Jaiswal AK (2019) Moving towards the second generation of lignocellulosic biorefineries in the EU: drivers, challenges, and opportunities. Renew Sust Energ Rev 101:590–599

Ho BKX, Azahari B, Yhaya MFB, Talebi A, Ng CWC, Tajarudin HA, Ismail N (2020) Green technology approach for reinforcement of calcium chloride cured sodium alginate films by isolated bacteria from palm oil mill effluent (POME). Sustainability 12:9468

Hongzhang C, Yumei W, Shuhua D (2011) Production of protein feed from sweet sorghum stalk by the two-step solid state fermentation. J Biofertil Biopestici 3:112

Hsu P-K, Liu C-P, Liu L-Y, Chang C-H, Yang S-S (2013) Protein enrichment and digestion improvement of napiergrass and pangolagrass with solid-state fermentation. J Microbiol Immunol Infect 46(3):171–179

Hu C-C, Liu L-Y, Yang S-S (2012) Protein enrichment, cellulase production and in vitro digestion improvement of pangolagrass with solid state fermentation. J Microbiol Immunol Infect 45(1): 7–14

Hu J, Xue Y, Guo H, Gao MT, Li J, Zhang S, Tsang YF (2017) Design and composition of synthetic fungal-bacterial microbial consortia that improve lignocellulolytic enzyme activity. Bioresour Technol 227:247–255

Huang LZ, Ma MG, Ji XX, Choi SE, Si C (2021) Recent developments and applications of hemicellulose from wheat straw: a review. Front Bioeng Biotechnol 9:1–14. https://doi.org/10.3389/fbioe.2021.690773

Iyer G, Chattoo BB (2003) Purification and characterization of laccase from the rice blast fungus, magnaporthegrisea. FEMS Microbiol Lett 227:121–126

Joshi A, Kashyap R, Khanna S (2011) Production of xylanase by Promicromonosporasp MARS with rice straw under non sterile conditions. Process Biochem 46(8):1614–1618. https://doi.org/10.1016/j.procbio.2011.05.003

Juturu V, Wu JC (2014) Microbial exo-xylanases: a mini review. Appl Biochem Biotechnol 174: 81–92. https://doi.org/10.1007/s12010-014-1042-8

Kanti A, Sudiana IM (2018) Production of phytase, amylase and cellulase by *Aspergillus*, *Rhizophus* and *Neurospora* on mixed rice straw powder and soybean curd residue. IOP Conf Ser Earth Environ Sci 166(1):012010. https://doi.org/10.1088/1755-1315/166/1/012010

Karthick Raja Namasivayam S, Babu M, Bharani RSA (2015) Evaluation of lignocellulosic agro wastes for the enhanced production of extracellular cellulase and xylanase by trichodermaharzianum. Nat Environ Pollut Technol 14:47–52

Kaur D, Bhardwaj NK, Lohchab RK (2017) Prospects of rice straw as a raw material for paper making. Waste Manag 60:127–139. https://doi.org/10.1016/J.WASMAN.2016.08.001

Ketsakhon P, Thammasittirong A, Thammasittirong S (2022) Adding value to rice straw waste for high-level xylanase production using a new isolate of Bacillus altitudinis RS3025. Folia Microbiol 68(1):87–99. https://doi.org/10.1007/s12223-022-00998-x

Khosravi A, Razavi SH (2020) The role of bioconversion processes to enhance bioaccessibility of polyphenols in rice. Food Biosci 35:100605. https://doi.org/10.1016/J.FBIO.2020.100605

Kogo T, Yoshida Y, Koganei K, Matsumoto H, Watanabe T, Ogihara J, Kasumi T (2017) Production of rice straw hydrolysis enzymes by the fungi Trichoderma reesei and Humicolainsolens using rice straw as a carbon source. Bioresour Technol 233:67–73. https://doi.org/10.1016/j.biortech.2017.01.075

Kubicek CP (2013) Systems biological approaches towards understanding cellulase production by Trichoderma reesei. J Biotechnol 163:133–142. https://doi.org/10.1016/j.jbiotec.2012.05.020

Kumar S (2015) Role of enzymes in fruit juice processing and its quality enhancement. Adv Appl Sci Res 6:114–124

Kumar S, Sharma HK, Sarkar BC (2011) Effect of substrate and fermentation conditions on pectinase and cellulase production by *Aspergillus Niger* NCIM 548 in submerged (SmF) and solid state fermentation (SSF). Food Sci Biotechnol 20:1289. https://doi.org/10.1007/s10068-011-0178-3

Kumar S, Paritosh K, Pareek N, Chawade A, Vivekanand V (2018) De-construction of major Indian cereal crop residues through chemical pretreatment for improved biogas production: an overview. Renew Sust Energ Rev 90:160–170. https://doi.org/10.1016/J.RSER.2018.03.049

Kumar S, D'Silva TC, Chandra R, Malik A, Vijay VK, Misra A (2021) Strategies for boosting biomethane production from rice straw: a systematic review. Bioresour Technol Rep 15:100813. https://doi.org/10.1016/j.biteb.2021.100813

Kumari B, Sudhakar P, Hemamalini K, Sree N, Pendyala V (2014) Studies on pectinase production by *Bacillus subtilis* using agro-industrial wastes. Res J Pharm Biol Chem Sci 5(6):330–339

Leah R, Kigel J, Svendsen I, Mundy J (1995) Biochemical and molecular characterization of a barley seed β-glucosidase. J Biol Chem 270:15789–15797. https://doi.org/10.1074/jbc.270.26.15789

Lee S, Kang M, Bae J-H, Sohn J-H, Sung BH (2019) Bacterial valorization of lignin: strains, enzymes, conversion pathways, biosensors, and perspectives. Front Bioeng Biotechnol 7:209

Li Q, Yi L, Marek P, Iverson BL (2013) Commercial proteases: present and future. FEBS Lett 587:1155–1163. https://doi.org/10.1016/j.febslet.2012.12.019

Li G, Liu X, Yuan L (2017) Improved laccase production by Funaliatrogii in absorbent fermentation with nutrient carrier. Biosci Bioeng 124(4):381–385. https://doi.org/10.1016/j.jbiosc.2017.05.002

Liao JJ, Latif NHA, Trache D, Brosse N, Hussin MH (2020) Current advancement on the isolation, characterization and application of lignin. Int J Biol Macromol 162:985–1024

López-Mondéjar R, Algora C, Baldrian P (2019) Lignocellulolytic systems of soil bacteria: a vast and diverse toolbox for biotechnological conversion processes. Biotechnol Adv 37:107374

Madhavi V, Lele SS (2009) Laccase: properties and applications. Bioresources 4:1694–1717

Maftukhah S, Abdullah A (2018) Cellulase enzyme production from rice straw using solid state fermentation and fungi Aspergillus niger ITBCC L74. MATEC Web Conf 156:01010. https://doi.org/10.1051/matecconf/201815601010

Mahalakshmi N, Jayalakshmi S (2016) Amylase, cellulase and xylanase production from a novel bacterial isolate achromobacter xylosoxidans isolated from marine environment. Int J Adv Res Biol Sci 3(1):230–233

Malgas S, Thoresen M, van Dyk JS, Pletschke BI (2017) Time dependence of enzyme synergism during the degradation of model and natural lignocellulosic substrates. Enzym Microb Technol 103:1–11

Mamma D, Hatzinikolaou DG, Christakopoulos P (2004) Biochemical and catalytic properties of two intracellular β-glucosidases from the fungus Penicillium decumbens active on flavonoid glucosides. J Mol Catal B Enzym 27:183–190

Mandal A (2015) Review on microbial xylanases and their applications. Int J Life Sci 4:178–187

Martínez ÁT, Ruiz-Dueñas FJ, Martínez MJ, Del Río JC, Gutiérrez A (2009) Enzymatic delignification of plant cell wall: from nature to mill. Curr Opin Biotechnol 20:348–357. https://doi.org/10.1016/j.copbio.2009.05.002

McCleary BV, Matheson NK (1983) Action patterns and substrate-binding requirements of β-d-mannanase with mannosaccharides and mannan-type polysaccharides. Carbohydr Res 119:191–219

McDonald JE, Rooks DJ, McCarthy AJ (2012) Methods for the isolation of cellulose-degrading microorganisms. Methods Enzymol 510:349–374

Meetei TT, Kundu MC, Devi YB (2020) Long-term effect of rice-based cropping systems on pools of soil organic carbon in farmer's field in hilly agroecosystem of Manipur, India. Environ Monit Assess 192:209

Meurial D, Uthandi S (2020) Xylanolytic Bacillus flexus PSX1.isolated from earthworm casts for xylanase production and biomass conversion. Madras Agric J 107:1–4. https://doi.org/10.29321/MAJ.S.000447

Mishra BK, Pandey Lata AK (2007) Lignocellulolytic enzyme production from submerged fermentation of paddy straw. Indian J Microbiol 47(2):176–179. https://doi.org/10.1007/s12088-007-0034-6

Mishra V, Jana AK, Maiti M, Gupta A (2017) Improvement of selective lignin degradation in the fungal pretreatment of sweet sorghum bagasse using synergistic CuSO 4-syringic acid supplements. J Environ Manage 193:558–566. https://doi.org/10.1016/j.jenvman.2017.02.057

Mohee R, Unmar GD, Mudhoo A, Khadoo P (2008) Biodegradability of biodegradable/degradable plastic materials under aerobic and anaerobic conditions. Waste Manag 28:1624–1629

Mondal S, Naik SK, Haris AA, Mishra JS, Mukherjee J, Rao KK, Bhatt BP (2020) Effect of conservation tillage and rice-based cropping systems on soil aggregation characteristics and carbon dynamics in Eastern Indo-Gangetic Plain. Paddy Water Environ 18:573–586

Morozova OV, Shumakovich GP, Shleev SV, Yaropolov YI (2007) Laccase-mediator systems and their applications: a review. Appl Biochem Microbiol 43(5):523–535. https://doi.org/10.1134/S0003683807050055

Muaaz-Us-Salam S, Cleall PJ, Harbottle MJ (2020) Application of enzymatic and bacterial biodelignification systems for enhanced breakdown of model lignocellulosic wastes. Sci Total Environ 728:138741

Munishamanna KB, Suresha KB, Veena R, Subramanya S (2017) Solid state fermentation of mango peel and mango seed waste by different yeasts and bacteria for nutritional improvement. Int J Food Ferment Technol 7(1):111

Muthulakshmi C, Gomathi D, Kumar DG, Ravikumar G, Kalaiselvi M, Uma C (2011) Production, purification and characterization of protease by Aspergillus flavus under solid state fermentation. Jordan J Biol Sci 4(3):137–148

Naga Padma P, Sravani P, Mishra PN, Sneha N, Anuradha K (2017) Synergestic effects of multiple enzymes on apple juice clarification. Ind J Sci Technol 10(10):1–5. https://doi.org/10.17485/ijst/2017/v10i10/107716

Naganagouda K, Salimath PV, Mulimani VH (2009) Purification and characterization of endo-beta-1,4 mannanase from Aspergillus niger gr for application in food processing industry. J Microbiol Biotechnol 19:1184–1190

Nagar R, Trivedi SK, Nagar D, Karnawat M (2020) Impact of agriculture crop residue burning on environment and soil health. Biotica Res Today 2:171–173

Naher L, Fatin SN, Sheikh MAH, Azeez LA, Siddiquee S, Zain NM, Karim SMR (2021) Cellulase enzyme production from filamentous fungi trichoderma reesei and aspergillusawamori in submerged fermentation with rice straw. J Fungi 7:868. https://doi.org/10.3390/jof7100868

Nair SG, Sindhu R, Shashidhar S (2008) Fungal xylanase production under solid state and submerged fermentation conditions. Afr J Microbiol Res 2:82–86

Nandan R, Poonia SP, Singh SS, Nath CP, Kumar V, Malik RK, McDonald A, Hazra KK (2021) Potential of conservation agriculture modules for energy conservation and sustainability of rice-based production systems of Indo-Gangetic Plain region. Environ Sci Pollut Res 28:246–261

Neelkant KS, Shankar K, Jayalakshmi SK, Sreeramulu K (2019) Optimization of conditions for the production of lignocellulolytic enzymesby Sphingobacterium sp. ksn-11 utilizing agro-wastes under submerged condition. Prep Biochem Biotechnol 49(9):927–934. https://doi.org/10.1080/10826068.2019.1643735

Niladevi KN, Sukumaran RK, Prema P (2007) Utilization of rice straw for laccase production by Streptomyces psammoticus in solid-state fermentation. J Ind Microbiol Biotechnol 34(10): 665–674. https://doi.org/10.1007/s10295-007-0239-z

Oboh G (2006) Nutrient enrichment of cassava peels using a mixed culture of saccharomyces cerevisiae and lactobacillus spp solid media fermentation techniques. Electron J Biotechnol 9(1):46–49

Orozco AL, Perez MI, Guevara O, Rodríguez J, Hernandez M, Gonzalez-Vila FJ, Polvillo O, Arias ME (2008) Biotechnological enhancement of coffee pulp residues by solid-state fermentation with streptomyces. Py-GC/MS analysis. J Anal Appl Pyrol 81(2):247–252

Pasha KM, Anuradha P, Subbarao D (2013) Applications of pectinases in industrial sector. J Pure Appl Sci Technol 16:89–95

Patel NS, Fung SM, Zanichelli A, Cicardi M, Cohn JR (2013) Ecallantide for treatment of acute attacks of acquired C1 esterase inhibitor deficiency. Allergy Asthma Proc 34(1):72–77. https://doi.org/10.2500/aap.2013.34.3620

Patela A, Divechab J, Shaha A (2021) Fomitopsismeliae CFA2, a novel brown rot for endoglucanase: emphasis towards enhanced endoglucanase production by statistical approach. Mycology 12(4):325–340. https://doi.org/10.1080/21501203.2021.1918277

Payne CM, Knott BC, Mayes HB, Hansson H, Himmel ME, Sandgren M et al (2015) Fungal cellulases. Chem Rev 115:1308–1348. https://doi.org/10.1021/cr500351c

Pedrolli DB, Monteiro AC, Gomes E, Cano CE (2009) Pectin and pectinases: production, characterization and industrial application of microbial pectinolytic enzymes. Open Biotechnol J 3:9–18. https://doi.org/10.2174/1874070700903010009

Qaseem MF, Shaheen H, Wu AM (2021) Cell wall hemicellulose for sustainable industrial utilization. Renew Sustain Energy Rev 144:110996. https://doi.org/10.1016/j.rser.2021.110996

Rahayu A, Utama P, Nurulita Y, Miranti M, Nugroho T (2019) Surfactant, nitrogen and carbon media optimization for trichoderma asperellum LBKURCC1 laccase production by flask solid state fermentation of rice straw. J Phys Conf Ser 1351(1):012030. https://doi.org/10.1088/1742-6596/1351/1/012030

Raheem A, Sajid M, Iqbal MS, Aslam H, Bilal M, Rafiq F (2019) Microbial inhabitants of agricultural land have potential to promote plant growth but they are liable to traditional practice of wheat (T. aestivum L) straw burning. Biocatal Agric Biotechnol 18:101060

Ralph J, Lundquist K, Brunow G, Lu FC, Kim H, Schatz PF, Marita JM, Hatfield RD, Ralph SA, Christensen JH, Boerjan W (2004) Lignins: natural polymers from oxidative coupling of 4-hydroxyphenyl- propanoids. Phytochem Rev 3:29–60

Ramos M, Laveriano E, Sebastian LS, Perez M, Jimenez A, Lamuela-Raventos RM, Garrigos MC, Vallverdú-Queralt A (2022) Rice straw as a valuable source of cellulose and polyphenols: applications in the food industry. Trends Food Sci Technol 131:14–27. https://doi.org/10.1016/j.tifs.2022.11.020

Rao MB, Tanksale AM, Ghatge MS, Deshpande VV (1998) Molecular and biotechnological aspects of microbial proteases. Microbiol Mol Biol Rev 62(3):597–635

Razali NAM, Mohd Sohaimi R, Othman RNIR, Abdullah N, Demon SZN, Jasmani L et al (2022) Comparative study on extraction of cellulose fiber from rice straw waste from chemo-mechanical and pulping method. Polymers 14(3):387. https://doi.org/10.3390/polym14030387

Rebello S, Anju M, Aneesh EM, Sindhu R, Binod P, Pandey A (2017) Recent advancements in the production and application of microbial pectinases: an overview. Rev Environ Sci Biotechnol 16:381–394. https://doi.org/10.1007/s11157-017-9437-y

Renge V, Khedkar S, Nandurkar NR (2012) Enzyme synthesis by fermentation method: a review. Sci Rev Chem Commun 2(4):585–590

Rodríguez Couto S, Ángeles Sanromán M (2006) Application of solid-state fermentation to food industry—a review. J Food Eng 76(3):291–302. https://doi.org/10.1016/j.jfoodeng.2005.05.022
Saadoun I, Dawagreh A, Jaradat Z, Ababneh Q (2013) Influence of culture conditions on pectinase production by Streptomyces sp. (strain J9). Int J Life Sci Med Res 3:148. https://doi.org/10.5963/LSMR0304002
Saanu AB, Oladiti OO (2018) Starter culture development from lactic acid bacteria for improved nutritive value of linamarase treated cassava peels. J Biomater Dent 2(1):1–6
Sadhasivam S, Savitha S, Swaminathan K, Lin FH (2008) Production, purification and characterization of mid-redox potential laccase from a newly isolated Trichoderma harzianum WL1. Process Biochem 43:736–742. https://doi.org/10.1016/j.procbio.2008.02.017
Salleh AB, Razak CNA, Rahman RNZRA, Basri M (2006) Protease introduction. In: Salleh AB, Rahman RNZRA, Basri M (eds) New lipases and proteases. Nova Science, New York, pp 23–39
Sanghi A, Garg N, Gupta VK, Mittal A, Kuhad RC (2010) One step purification and characterization of a cellulose free xylanase produced by alkalophilic Bacillus subtilis ASH. Braz J Microbiol 41:467–476. https://doi.org/10.1590/S151783822010000200029
Saranraj P (2011) Optimization for cellulase production by Aspergillus niger using paddy straw as substrate. Int J Adv Sci Tech Res 1(1):68–85
Saratale GD, Saratale RG, Ghodake GS, Jiang YY, Chang JS, Shin HS, Kumar G (2017) Solid state fermentative lignocellulolytic enzymes production, characterization and its application in the saccharification of rice waste biomass for ethanol production: an integrated biotechnological approach. J Taiwan Inst Chem Eng 76:51–58
Sarkar N, Aikat K (2014) Aspergillus fumigatus NITDGPKA3 provides for increased cellulase production. Int J Chem Eng 2014:959845. https://doi.org/10.1155/2014/959845
Schülein M (1988) Cellulases of trichoderma reesei. Methods Enzymol 160:234–242. https://doi.org/10.1016/0076-6879(88)60125-X
Sharifzadeh F, Dizajnejad G, Mohammadhassanpour S, Rahnama N (2020) Cellulase production by Penicillium expansum MDFS2 by solid-state fermentation of rice straw, rice bran, and wheat straw residues. Adv Res Microb Metabol Technol 2:101–108. https://doi.org/10.22104/armmt.2020.4326.1045
Sharma PK, Chand D (2012) Production of cellulase free thermostable xylanase from pseudomonas sp. XPB-6. Int Res J Biol Sci 1(5):31–41
Sindhu R, Binod P, Pandey A (2016) α-Amylases. In: Pandey A, Negi S, Soccol CR (eds) Current developments in biotechnology and bioengineering—production, isolation and purification of industrial products. Elsevier, Amsterdam, pp 3–24. https://doi.org/10.1016/B978-0-444-63662-1.00001-4
Sindhu R, Binod P, Madhavan A, Beevi US, Mathew AK, Abraham A, Pandey A, Kumar V (2017) Molecular improvements in microbial α-amylases for enhanced stability and catalytic efficiency. Bioresour Technol 245(Part B):1740–1748. https://doi.org/10.1016/j.biortech.2017.04.098
Singh L, Brar BS (2021) A review on rice straw management strategies. Nat Environ Pollut Technol 20(4):1485–1493. https://doi.org/10.46488/NEPT.2021.v20i04.010
Singh RB, Saha RC, Singh M, Chandra D, Shukla SG, Walli M, Pradhan PK, Kessels HPP (1995) Rice straw—its production and utilization in India. ICAR, New Delhi
Singh R, Mittal A, Kumar M, Mehta PK (2016) Microbial protease in commercial applications. J Pharm Chem Biol Sci 4(3):365–374
Singh P, Singh G, Sodhi GPS (2020a) Energy and carbon footprints of wheat establishment following different rice residue management strategies vis-à-vis conventional tillage coupled with rice residue burning in North-Western India. Energy 200:117554
Singh R, Yadav DB, Ravisankar N, Yadav A, Singh H (2020b) Crop residue management in rice-wheat cropping system for resource conservation and environmental protection in North-Western India. Environ Dev Sustain 22:3871–3896

Singh G, Gupta MK, Chaurasiya S, Sharma VS, Pimenov DY (2021) Rice straw burning: a review on its global prevalence and the sustainable alternatives for its effective mitigation. Environ Sci Pollut Res 28(25):32125–32155. https://doi.org/10.1007/s11356-021-14163-3

Singhvi MS, Gokhale DV (2019) Lignocellulosic biomass: hurdles and challenges in its valorization. Appl Microbiol Biotechnol 103:9305–9320

Subramaniyam R, Vimala R (2012) Solid state and submerged fermentation for the production of bioactive substances: a comparative study. Int J Sci Nat 3(3):480–486

Suganthi R et al (2011) Amylase production by Aspergillus Niger under solid state fermentation using agroindustrial wastes. Int J Eng Sci Technol 3(2):1756–1763

Sukumaran RK (2009) Bioethanol from lignocellulosic biomass, part II. Production of cellulases and hemicellulases. In: Pandey A (ed) Handbook of plant-based biofuels. CRC Press, Boca Raton, FL, pp 141–157

Sukumaran RK, Singhania RR, Pandey A (2005) Microbial cellulases—production, applications and challenges. J Sci Ind Res 64:832–844

Sun J, Zheng M, Lu Z, Lu F, Zhang C (2017) Heterologous production of a temperature and pH-stable laccase from Bacillus vallismortis FMB-103 in Escherichia coli and its application. Process Biochem 55:77–84. https://doi.org/10.1016/j.procbio.2017.01.030

Sweeney MD, Xu F (2012) Biomass converting enzymes as industrial biocatalysts for fuels and chemicals: recent developments. Catalysts 2:244–263. https://doi.org/10.3390/catal2020244

Taha M, Kadali KK, Khalid AH, Smith AT, Ball AS, Adetutu EM (2015a) An effective microplate method (biolog MT2) for screening native lignocellulosic-straw-degrading bacteria. Ann Microbiol 65:2053–2064

Taha M, Shahsavari E, Al-Hothaly K, Mouradov A, Smith AT, Ball AS, Adetutu EM (2015b) Enhanced biological straw saccharification through coculturing of lignocellulose-degrading microorganisms. Appl Biochem Biotechnol 175:3709–3728

Tamilanban R, Velayudhan SS, Ebenezer S, Harshavardhan S (2017) Purification and characterization of an extracellular cellulase produced using alkali pretreated rice straw by stenotrophomonas maltophilia. Int J Biol Res 2(4):45–54

Taneja G, Pal BD, Joshi PK, Aggarwal PK, Tyagi NK (2019) Farmers' preferences for climate-smart agriculture—an assessment in the Indo-Gangetic plain. In: Pal BD, Kishore A, Joshi PK, Tyagi NK (eds) Climate smart agriculture in South Asia. Springer, Singapore, pp 91–111

Tonin F, Melis R, Cordes A, Sanchez-Amat A, Pollegioni L, Rosini E (2016) Comparison of different microbial laccases as tools for industrial uses. New Biotechnol 33(3):387–398. https://doi.org/10.1016/j.nbt.2016.01.007

Torregrosa A, Miguel Giner J, Velazquez-Martí B (2021) Equipment performance, costs and constraints of packaging and transporting rice straw for alternative uses to burning in the "parc natural l'albufera de val'encia" (Spain). Agriculture 11(6):570. https://doi.org/10.3390/agriculture11060570

Trivedi A, Verma AR, Kaur S, Jha B, Vijay V, Chandra R, Vijay VK, Subbarao PMV, Tiwari R, Hariprasad P, Prasad R (2017) Sustainable bio-energy production models for eradicating open field burning of paddy straw in Punjab, India. Energy 127:310–317

Tsegaye B, Balomajumder C, Roy P (2019) Microbial delignification and hydrolysis of lignocellulosic biomass to enhance biofuel production: an overview and future prospect. Bull Natl Res Cent 43:51. https://doi.org/10.1186/s42269-019-0094-x

Urfels A, McDonald AJ, Krupnik TJ, Van Oel PR (2020) Drivers of groundwater utilization in water-limited rice production systems in Nepal. Water Int 45:39–59

Van der Maarel MJEC, van der Veen B, Uitdehaag JCM, Leemhuis H, Dijkhuizen L (2002) Properties and applications of starch-converting enzymes of the α-amylase family. J Biotechnol 94:137–155. https://doi.org/10.1016/S0168-1656(01)00407-2

Van Hung N, Maguyon-Detras MC, Migo MV, Quilloy R, Balingbing C, Chivenge P, Gummert M (2020) Rice straw overview: availability, properties, and management practices. In: Gummert M, Hung N, Chivenge P, Douthwaite B (eds) Sustainable rice straw management. Springer, Cham, pp 1–13. https://doi.org/10.1007/978-3-030-32373-8_1

Vijaya C, Singaracharya MA (2005) Delignification of paddy straw by a white rot fungus, pleurotus ostrreatus. Asian J Microbiol Biotechnol Environ Sci 7(1):77–83

Wi S, Choi I, Kim K, Kim H, Bae HJ (2013) Bioethanol production from rice straw by popping pretreatment. Biotechnol Biofuels 6:166

Xie S, Ragauskas AJ, Yuan JS (2016) Lignin conversion: opportunities and challenges for the integrated biorefinery. Ind Biotechnol 12:161–167

Yuliana T, Putri NZ, Komara DZ, Mardawati E, Lanti I, Rahimah S (2020) Study of ganodermalucidum in laccase production using corncob and paddies straw substrates on submerged fermentation system. Pak J Biol Sci 23:1060–1065. https://doi.org/10.3923/pjbs.2020.1060.1065

Zabidi NAM, Foo HL, Loh TC, Mohamad R, Rahim RA (2020) Enhancement of versatile extracellular cellulolytic and hemicellulolytic enzyme productions by lactobacillus plantarum RI 11 isolated from Malaysian food using renewable natural polymers. Molecules 25:2607

Zahari N, Md Shah UK, Mohamed Asa'ari AZ, Mohamad R (2016) Selection of potential fungi for production of cellulase-poor xylanase from rice straw. Bioresources 11(1):1162–1175. https://doi.org/10.15376/biores.11.1.1162-1175

Zheng G, Yin T, Lu Z, Yannick Benz Boboua S, Li J, Zhou W (2020) Degradation of rice straw at low temperature using a novel microbial consortium LTF-27 with efficient ability. Bioresour Technol 304:123064

Zhiqiang L, Xu A, Zhao T (2011) Energy from the combustion of rice straw: status and challenges to China. Energy Power Eng 3:325–331

Chapter 10
Biochar: A Pyrolyzed Green Fuel from Paddy Straw

Tanvi Sahni, Diksha Verma, and Sachin Kumar

Abstract Rice straw is a major by-product that possess significant environmental challenges due to its increasing production and the prevalent practice of open burning for disposal. This practice not only leads to air pollution but also hinders soil fertilization and crop establishment. To address these issues, the conversion of rice straw into biochar has emerged as a potential solution. Biochar is produced from rice straw through a thermochemical conversion known as pyrolysis. Due to the environmental benefits and energy efficiency of biochar, it has acquired great recognization. However, concerns regarding the greenhouse gas (GHG) mitigation potential of biochar production need to be addressed. This chapter provides insights into the synthesis of biochar using agricultural waste. RSB (rice straw biochar) is considered a suitable amendment for the ecological and sustainable development of ecosystems. It offers several advantages, including carbon neutrality, slow-release effects of nutrients, and improved soil water content and porosity. Additionally, RSB has been found effective in wastewater and soil contamination treatment. It acts as a sorbent for inorganic and organic sorbents due to its cation exchange ability and wide surface area. The physical properties of rice straw, such as bulk density and moisture content, plays an important function in its handling as well as storage. Processing methods such as pelletization and briquetting can increase its density and reduce its volume for easier storage and transportation. The thermal properties of rice straw, including calorific value and volatile matter, influence its conversion into biofuels. Chemically, cellulose, hemicellulose, and lignin are major components of lignocellulose found abundantly in biochar. These polymers vary quantitatively and qualitatively based on factors such as variety, season, and geographical location. Understanding the rice straw's chemical composition is essential for the utilization of it as a feedstock and soil fertility enhancer. The biochar production from rice straw is carried out through thermochemical conversion technologies, including pyrolysis. However, the GHG mitigation potential of biochar production needs further

T. Sahni · D. Verma
Punjab Agricultural University, Ludhiana, Punjab, India

S. Kumar (✉)
Food Corporation of India, Ludhiana, Punjab, India

N. Srivastava et al. (eds.), *Paddy Straw Waste for Biorefinery Applications*, Clean Energy Production Technologies, https://doi.org/10.1007/978-981-99-8224-0_10

investigation to optimize the process and enhance its environmental benefits. In nutshell, the utilization of rice straw as biochar offers a sustainable and efficient way to manage agricultural waste, improve soil fertility, and mitigate environmental issues. Further research and development are required to optimize the production process and explore the full potential of rice straw biochar in various applications.

Keywords Biochar · Rice straw · Pyrolysis · Rice straw biochar · Green fuel

10.1 Introduction

Over 50% of the earth's population hinge on rice as a staple crop, and 70–80% (600–800 t) is produced from 143 million hectares (Mha) in Asia (Ministry of Agriculture and Farmers Welfare 2021). In Asia, India possessed first rank for the area under rice cultivation (44.6 Mha) but was followed by China in rice production (Rani and Paul 2023). Most of India's rice cropping land is located on the Gangetic and Punjab plains where 124 million metric tons rice was produced during 2021–2022 in states such as Punjab, Uttar Pradesh, West Bengal, etc. Rice straw is a major by-product formed during the harvesting of rice. In harvested crops, except grain, all the parts (stems, leaves, and spikelets) comprise rice straw (Rathour et al. 2023). This rate is increasing alarmingly due to shorter turnaround crop time and as a consequence leads to open burning for its disposal. Open burning creates oblivion in estimating rice straw production and also causes air pollution through the emission of CO_2, CO, CH_4, NO_x, and SO_x gases (Gadde et al. 2009a, 2009b; Mendoza and Samson 1999). In intensive cropping systems, rice straw is usually grown two to three times a year, which provides a shorter period for decomposition (Palaniappan 2006). This results in poor soil fertilization and crop establishment. Rice straw is also used for mushroom plantation, activated charcoal, biodiesel, and ethanol production (Goodman 2020).

Due to its importance in agriculture, rice straw-derived biochar has gained attentiveness (Xu et al. 2020; Warnock et al. 2007; Vu and Nguyen 2011). The optimum temperature range for RSD production through pyrolysis was found to be 300–600 °C (Medha et al. 2023). It brings carbon neutrality to the agricultural system, but one of the major concerns toward biochar production is their corresponding GHG (greenhouse gases) mitigation potential (Xia et al. 2023). According to the 2018 Chinese census, 145 Mt/year of biochar are produced from different waste sources, representing GHG emissions. To mitigate GHG emissions, biochar should be prepared from crop residues, livestock manure, and food waste at pyrolysis temperatures of 300–500 °C (Xia et al. 2023). Adsorption and partitioning are commonly investigated mechanisms for their reported potential, and here biochar possesses wide surface area and CEC ratio (Ahmad et al. 2014; Cao et al. 2011). Biochar is structurally found similar to activated charcoal. These two materials are prepared by pyrolysis, and they are both used as sorbents to control contaminants in water and soil that includes inorganic and organic sludges (Enaime et al. 2020).

Fig. 10.1 Rice to biochar

According to Ibrahim et al. (2016), rice husk biochar (RHB) is a beneficial soil amendment for contaminated soils. Lv et al. (2023) reported the use of RSB in rice-based cropping systems which results in the slow release of ammoniacal nitrogen and regulates the nitrogen uptake microbially. In one of the studies, an observation was noticed that when rice husk biochar and straw compost was applied for the second time, it caused hike in total carbon and exchangeable potassium in the upland rice and soybean cropping system. Along with this, biochar also provides significant improvement in soil water content: total porosity, micro-pores, and soybean yield (Barus et al. 2023). Biochar also helps in wastewater treatment and the removal of soil contaminants (Dai et al. 2020). Copper ions, one of the leading soil contaminants, were adsorbed by RSB in a batch experiment study by Mei et al. (2020). It also provides reactive sites for microbial action (Li et al. 2019). Thus, RSB is found as a suitable amendment for the efficient, ecological, and sustainable development of the ecosystem. The basic mechanism of RSB production is given below (Fig. 10.1).

10.2 Characteristics of Rice Straw

All living cells depend on atmospheric carbon dioxide for their growth. In rice straw, the concentrations of carbon, nitrogen, P_2O_5, K_2O, and sulfur were (40–47%), (0.5–0.8%), (0.07–0.12%), (1.16–1.66%), and (0.05–0.1%), respectively (Dobermann and Fairhurst 2002; Nghi et al. 2020). Hence, biomass is known as a carbon neutral fuel precursor (Ahamer 2022). It has low sulfur and nitrogen content as compared to solid coal. Hence, carbon dioxide is more recoverable during biomass burning as compared to coal (Demirbas 2004). In the process of biofuel production, biomass physical properties, thermal properties, and chemical composition are mainly considered for thermal conversion.

10.2.1 Physical Properties

Physical properties are mainly considered for the handling and storage of rice straws. It includes bulk density and moisture content. Bulk density varies from 13–18 kg/m^3 in dry matter (dm) of loose rice straw (Migo 2019) to 50–120 kg/m^3 in chopped rice straw of length 2–10 mm (Liu et al. 2011). It varies from equipment to equipment. The density of rice husks varies between 84 and 114 kg/m^3. Thus, loose rice straw has higher volume and lower density per kilogram. Thus, it is recommended to process rice straw via pellet mills (Pandey et al. 2023), roller presses (Das et al. 2023), piston presses (Wang et al. 2023), cubers (Abitha et al. 2023), briquette presses (Rathour et al. 2023), screw extruders (Cui et al. 2023), and agglomerators (Zeleke et al. 2023) to reduce rice straw volume and increase its density. Moisture content refers to the amount of water (%) present in a particular material. Rice straw's moisture level is crucial in determining how it will be processed and what it will be used for. A straw's moisture content can have an impact on its heating value, which is a factor that should be considered when the waste is intended to be used as a source of energy. Additionally, rice straw should be moist between 12% and 17% before compressing if it is to be reduced in volume (Kargbo et al. 2010).

10.2.2 Thermal Properties

The efficiency of biofuels produced from biomass was also evaluated by thermal properties. The calorific value is one of the thermal properties which express the amount of heat that can be produced by burning that fuel completely. CV of rice straw (14.6 MJ/kg) is one-third that of kerosene (46.2 MJ/kg). In terms of rice straw, we often expressed it in terms of heating value more conveniently than CV. This is further categorized into high heating value (HHV) and (LHV) low heating value. Rice straw's HHV varies in the range of 14.08–15.09 MJ/kg (Van Hung et al. 2020). Volatile matter (VOM) refers to material that gets converted to a gaseous state once the right conditions are met. Rice straw is characterized by high VOM (60.55–69.70%) that leads to easier burning conditions; this also creates difficulty in causing controlled combustion conditions (Liu et al. 2011).

10.2.3 Chemical Properties

Agricultural biomass rice straw is composed of lignocellulose which quantitatively and qualitatively varies with geographical locations, climatic conditions, season, and variety. The chemical composition provides its significance as livestock feed and soil fertility. Thus, the chemical composition of rice straw along with the ultimate element analysis from different countries is mentioned in Table 10.1.

Table 10.1 Chemical configuration of rice straw

S.No.	Components	Percentage (%) (dry matter basis)
1.	Total ash	18
2.	Crude protein	4
3.	Crude fiber	37
4.	Organic matter	82
5.	Non-fatty ester	43
6.	Neutral detergent fiber	75
7.	Acid detergent fiber	54
8.	Silica	8
9.	Cellulose	37
10.	Lignin	8
11.	Nitrogen	0.7
12.	Phosphate	0.23
13.	Potassium	1.75
14.	Sulfur	0.08
15.	Calcium	3.5
16.	Magnesium	2
17.	Iron oxide	0.85
18.	Titanium oxide	0.09
19.	Aluminum oxide	1.10

Moreover, rice straw is rich in cellulose (30–45%), hemicellulose (20–25%), lignin (15–20%), and ash content (8.5–20.4%) (Binod et al. 2010; Takkellapati et al. 2018), while all other polymers are present in smaller amounts. It is an amalgam of the lignin-phenolics-carbohydrate complex that provides robustness to cell wall. Rice straw is dominated by glucose (41–43.4%) and pentose sugars like xylose (14.8–20.2%) (Maiorella 1985; Roberto et al. 2003), along with others like arabinose (2.4–4.5%), galactose (0.4%), and mannose (1.8%) (Kumar et al. 2023) (Fig. 10.2). It contains a higher amount of esterified and etherified forms of ferulic acid and *p*-coumaric acid (Fig. 10.2).

10.2.3.1 Cellulose

D-Glucose subunits joined together to form cellulose. These are interlinked by inter- and intra-hydrogen bonding and β-1,4-glycosidic linkage to form a linear condense polymer known as cellulose. In rice straw, cellulose is crystalline in structure with a higher degree of polymerization (1820) which will provide mechanical strength to the cell wall (Yang et al. 2011; Abe and Yano 2009; Hallac and Ragauskas 2011).

10.2.3.2 Hemicellulose

Rice straw has xylan hemicellulose; it is the most reactive form of three polymers of lignocellulose which carbonize and are volatile at temperatures below 250 °C. It has

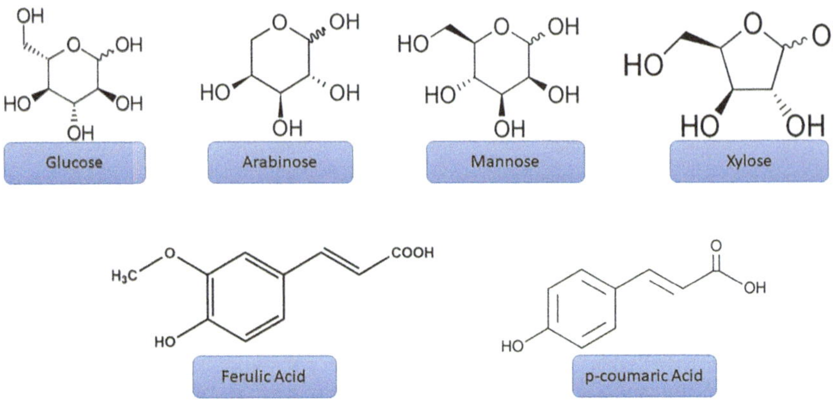

Fig. 10.2 Major compounds in rice straw

an average weight molecular weight of 18,800–48,700 and an average number molecular weight of 8200–15,900. It has a lower arabinose/xylose ratio (0.17) which means a higher degree of polymerization in the range of 80–200 °C (Kausar et al. 2016) which means a lower arabinose/xylose ratio, i.e., 0.17 (Bezerra and Ragauskas 2016). It also has a high degree of polymolecularity, poly-diversity, and polydispersity. Xylan which is a heteropolysaccharide contains aliphatic acids like acetic acid, aromatic acids like ferulic acid, p-coumaric acid, glucuronic acid, and its o-methyl ether too or sugars like arabinose. Their structures are drawn in Fig. 10.2. The subunits shown above have D-xylopyranose units in their backbone which are joined by 1,4-linkage, viz., glycosidic bonding. The connection between cellulose and hemicellulose is via hydrogen bonding, while hemicellulose and lignin are via covalent and ester bonding (Bezerra and Ragauskas 2016). Isolation and characterization of hemicellulose were done by hydrogen peroxide method, through alkali, alcohols, and a combination of alcohol/alkali hydrolysis through cleavage of the ester linkage.

10.2.3.3 Lignin

In general **lignin** in biomass consists of aromatic subunits, namely, methoxylated phenyl propanoic subunits. These subunits are similar to aromatic compounds in fuels (Shuai et al. 2016). Rice straw lignin is a GSH type where G stands for guaiacyl, S stands for syringyl, and H stands for p-hydroxy phenyl subunits as shown in Fig. 10.1. It also had β-O-4 linkage in alkyl aryl ether, β-5 linkage in phenyl coumarin compound (dehydrodiisoeugenol), β-β, α-O-4, and a minor amount of α,β-diarylether as shown in Fig. 10.3. Various physical, chemical, and microbial treatments like enzyme/mild acidolysis lignin method (EMAL) and spectroscopic studies like FT-IR-XRD, ^{13}C-NMR, ^{1}H-NMR, ^{31}P-NMR, and GC-MS spectral

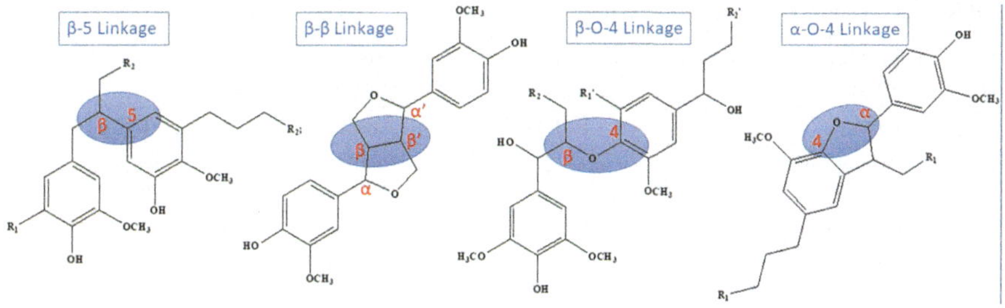

Fig. 10.3 Type of linkages in lignin

analysis were performed to investigate its structure. X-ray fluorescence study revealed that ash mainly consists of silica oxide (69.2%), calcium oxide (3.46%), potassium oxide (6.4%), magnesium oxide (2.81%), aluminum oxide (5.3%), sodium oxide (3.43%), and iron oxide (0.9%) (Morsy et al. 2022).

10.3 Production of Biochar

Bioenergy is an intermittent-free and stable source of energy with its ability to store biofuels from biomass in the character of solids, liquids, and gases. Biofuels are the fourth largest source of energy. They are the best alternative to decline anthropogenic CO_2 emissions, suppress deteriorated greenhouse gases, and mitigate global warming. Four main technologies are used to convert biomass into a green fuel: (i) physical, (ii) chemical, (iii) biological, and (iv) thermochemical. Physical conversion includes the drying of biomass, the separation of biomass, pulverization of biomass, and pelletization of biomass. Biodiesel production through transesterification is a notable chemical conversion. Biological conversions include biogas, biohydrogen, bioethanol, and biobutanol through fermentation, saccharification, or photosynthesis. It is noteworthy that thermochemical conversion is the most significant pathway for the production of biofuels, except for combustion, due to the production of heat as an end product. A variety of waste and blends can be processed with thermochemical conversion due to its efficient nutrient recovery, small footprint, and quick reaction time. Biochar can also be produced most efficiently using this method as cited in the literature (Table 10.2).

Production of biochar mainly proceeds through thermochemical conversion. Economically, four techniques are found operational with most of the researchers. These are (i) pyrolysis, (ii) hydrothermal carbonization, (iii) gasification, and (iv) torrefaction (Pang 2019; Lin et al. 2016). Depending upon the type of biomass, the abovementioned techniques were employed by optimizing conditions like heating rate, temperature, residence time, etc. For RSB production, all techniques were used wisely, and their instances were described in each section. The basic reactions occurring during these techniques are dehydration around 100 °C,

Table 10.2 Thermochemical conversion methods

S. No.	Thermochemical conversion methods	Temperature range (°C)	Oxygen supply	References
1.	Pyrolysis	400–800	Absent	Osayi et al. (2014)
2.	Gasification	600–1300	Insufficient	Heidenreich and Foscolo (2015)
3.	Torrefaction	200–300	Absent	Stępień et al. (2017)
4.	Liquefaction	200–600	Absent	Nkosi et al. (2021)
5.	Combustion	900–1500	Sufficient	Shanmugam et al. (2022)
6.	Carbonization	400–1000	Absent	Seow et al. (2022)

followed by degradation of lignin, cellulose, and hemicelluloses above 220 °C. The other by-products obtained during these methods are syngas and bio-oil (Abdelaziz et al. 2016).

10.3.1 Pyrolysis

Pyrolysis involves the anaerobic decomposition of biomass within 250–900 °C (Osayi et al. 2014). Mainly depolymerization, fragmentation, and cross-linking reactions take place during this process at specific temperatures. The gaseous emission involves syngas (C_1–C_2 hydrocarbons), hydrogen, carbon monoxide, and carbon dioxide. Commonly used reactors for pyrolysis are paddle kilns, wagons, agitated sand rotating kilns, and bubbling fluidized beds. In biochar production, the optimum temperature is key for the maximum yield of biochar, any increase after the optimum temperature will elevates the production of syngas and a decrease in product yield (Wei et al. 2019). Based on pressure, temperature, and rate of heating, pyrolysis is divided into i) slow pyrolysis and ii) fast pyrolysis.

10.3.1.1 Slow Pyrolysis

It involves pyrolysis at temperature of 5–7 °C with a longer residence time. This technique was found to be best in terms of biochar yield for production (Liu et al. 2015; Al Arni 2018).

10.3.1.2 Fast Pyrolysis

It involves moderate temperature pyrolysis (400–600 °C) for a shorter period (0.5–2 s) (Wang et al. 2014). In this process, applied energy converted solid biomass to biogas, biochar, and bio-oil among which bio-oil is the major end product. RSB is prepared by the solvothermal method at two optimum temperatures of 400 and 600 °C. Biochar thus obtained is combined with bentonite clay for reducing the leaching of phosphate and ammoniacal ions (Medha et al. 2023).

10.3.1.3 Catalytic Pyrolysis

This process proceeds at 350–650 °C with a sweep gas flow rate of 0.02–11.00 L min^{-1}. The most commonly used catalysts are zeolites (SiO Al_2O_3, HZSM-5, ZSM-5), modified zeolites (Ni/ZSM-5, Zn/ZSM-5, Ga/ZSM-5), metal oxide catalysts (i.e., ZnO, MgO, CaO), etc.

10.3.1.4 Microwave Pyrolysis

During this process, the disintegration occurs at 250–800 °C with sweep gas flow rates of 0.05–20.00 L/min for 5–60 min.

Reactions Occurring in Paddy Straw Biochar During Pyrolysis: Cellulose Decomposition

Cellulose degradation involves a reduction in the degree of polymerization. An intermediate formed during fast pyrolysis includes levoglucosan. It also undergoes dehydration to form hydroxyl-methyl furfural which underwent aromatization, condensation, and polymerization resulting in biochar production.

Hemicellulose Decomposition

Hemicellulose decomposition leads to the formation of oligosaccharides by depolymerization. It also involves decarboxylation, intramolecular rearrangement, depolymerization, and aromatization (Huang et al. 2012).

Lignin Decomposition

Lignin is decomposed at higher temperatures. The β-O-4 linkage as shown in the diagrams above was broken down to form free radicals. These radicals further react with protons present in the surrounding as derived from other species. Thus, a chain propagation reaction starts in this decomposition process (Mu et al. 2013).

10.3.2 Hydrothermal Carbonization

It is a profitable method in which heating is done with temperature range of 180–250 °C (Lee et al. 2018). The solid product procured after gasification and pyrolysis is known as biochar, while the hydrothermal carbonization product is known as hydrochar (Fang et al. 2018). In a closed reactor, rice straw is blended with water, and the temperature will increase gradually. Biochar was obtained below 250 °C, and the process is known as hydrothermal carbonization (Zhang et al. 2017). Bio-oil was obtained within 250–400 °C known as hydrothermal liquefaction. The products such as CH_4, H_2, CO_2, and CO, in their gaseous state, were obtained above 400 °C known as hydrothermal gasification (Khorram et al. 2016). Reactions that take place during heating are condensation, polymerization, and intramolecular dehydration, along with this products like 5-hydroxymethylfurfural, and their corresponding derivatives will be obtained (Bakraoui et al. 2020). Lignin is a complex matrix decomposed through reactions like dealkylation and hydrolysis. Phenol products like catechols and syringols were obtained during this decomposition (Jain et al. 2016). Hydrochar was obtained at last through re-polymerization and cross-linking which is much similar to the pyrolysis method.

10.3.3 Gasification

The aerobic combustion of biomass is commonly known as gasification. Gasification agents like oxygen, air, steam, etc. are used to decompose biomass. Depending upon the reaction temperature, two types of products were obtained, i.e., syngas and biochar. (i) Drying: this process involves drying if biomass is higher in moisture content without energy recovery. (ii) Combustion/oxidation: the second step is combustions/oxidation where gasifying agents like air, oxygen, sub-critical water, supercritical water, etc. react with combustible species to form products. Conventionally, gasification is classified into steam, conventional, and supercritical water gasification (Heidenreich and Foscolo 2015).

10.3.3.1 Conventional Gasification

In this method, the temperature ranges between 400 and 1200 °C using gasifying agent (air) with an equivalence ratio (ER) of 0.15–0.4 and pressure ranges of 0.1–1 MPa. Output carbon conversion (%) is 63–95, while cold gas efficiency (%) is 53–75.

10.3.3.2 Steam Gasification

In this method, the temperature ranges between 600 and 1000 °C using gasifying agent (steam) with an equivalence ratio (ER) of 0–0.37 and pressure ranges of 0.1–1 MPa. Output carbon conversion (%) is 42–124, while cold gas efficiency (%) is 33–80.

10.3.3.3 Supercritical Water Gasification

In this method, the temperature ranges between 374 and 800 °C using water as gasifying agent with pressure ranges of 22–50 MPa. Output carbon conversion (%) is 4–97, while cold gas efficiency (%) is 32–91.

Syngas was produced in higher amounts than biochar. It is comprised of H_2, CO, CO, and traces of hydrocarbons. Increase in temperature CO_2 and H_2 production increases, while CH_4, CO, and H_2 production decreases (Prabakar et al. 2018). During gasification, the biomass underwent reactions like water-gas reaction, complete oxidation, steam methane reforming, partial oxidation, water-gas shift reaction, water-gas reaction, and hydrocarbon reaction (Agarwal 2014; Young 2010). Researchers mainly aim to use these syngas as another efficient fuel for heating, drying, cooking, biofuel production, and electricity production. This is achieved by removing non-combustible gases and water from production. In one study from Young (2010), production of industrial chemical compounds like methanol, hydrogen, and ammonia can produce from CO and H_2 of syngas. Four types of gasifiers used in gasification are (i) fluidized bed, (ii) fixed bed, (iii) moving bed, and (iv) entrained-flow. Recent advancement in gasifiers is reported by Ngamchompoo (2018). In this research work, the characteristic of gasification of air of rice straw was evaluated in cyclone gasifiers. It was found that as gasifier temperature increases, equivalence ratio (ER) also increases which leads to improving gas quality and improved gasification performance. A smaller feedstock size with optimum moisture content is more favorable. Wang et al. (2020) studied air gasification with two-staging systems, i.e., medium-temperature bubbling fluidized gasifier and high-temperature swirl-flow gasifier in a pilot scale study using rice straw as biomass fuel. It was found that as gasification temperature increases, it improves the gasification efficiency, syngas quality, and carbon conversion with an increase in agglomeration. An increase in ER has declined syngas heating value, carbon conversion, tar yield, and tar concentration.

10.3.4 Torrefaction

Biochar is also produced through torrefaction, which is a relatively new technology. It employs a low heating rate, known as high temperature drying, roasting, slow and mild pyrolysis, and wood cooking. A variety of decomposition processes are used to remove the CO_2, O_2, and moisture present in biomass in the presence of inert atmospheric air and absence of oxygen under 200–300 °C for 15–60 min. These conditions should vary with different categories of biomass to accelerate the synthesis (Stępień et al. 2017). In torrefaction, based on substrate heating, reactors are mainly divided into two groups: direct and indirect. Indirect heating is found accompanied by auger and rotary reactors, while direct heating is accompanied by auger, microwave, moving bed, vibrating belt, multiple zone, rotary drum, and auger reactors. Usually, direct heating reactor works in anaerobic condition have a heating medium (Klimiuk et al. 2012; Koppejan et al. 2012). Rotary technologies mentioned earlier have been put into application successfully on a commercial scale in the Netherlands, the USA, Canada, Sweden, Germany, Spain, and France (Nhuchhen et al. 2014).

10.4 Applications of RSB

A pot experiment (2 years) was done for the validation of the impact of rice straw with varying combinations (5) of biochars on rice crops. Only an appropriate amount of biochar, viz., 20 t/ha, with straw 7 t/ha was found to provide authentic results in terms of alleviating N and P losses, increasing rice growth and yield, phosphorous availability, and nitrogen retention (Li et al. 2023).

Not only biochar, their modified versions like rice husk-derived silicon-rich biochar and their iron-rich modified versions were implemented to improve soil's various characteristics to lighten the PTE accumulation in rice cultivated pot trials. RSB was employed for the removal of ciprofloxacin (CIP) antibiotic commonly found in river sediments, domestic sewage, and pharmaceutical wastewater. The adsorption mechanism of RSB was found maximum with pH = 5; maximum adsorption capacity is 747.64 mg/g through the carbonyl (C=O) bond. It also revealed the role of electrostatic interactions, π-π interactions, and hydrogen bond interactions. Functionalization of biochar surface with TiO_2 increases the photocatalytic performance and degradation of CIP (Qu et al. 2023).

Sakhiya et al. (2023) reported heavy metal removal from groundwater through RSB. Biochar was produced at different temperatures, viz., 400–600 °C, through pyrolysis. RSB produced at 600 °C was found suitable for Manganese and arsenic removal from water due to wide surface area and adsorption capability. Electrostatic attractions between metalloid and surface functional groups like COOH and OH facilitate adsorption. Biochar applied at the rate of 0.1 g in 50 mL water had the highest removal efficiency and lowers cancer risks in the community.

Biochar application may possess adverse effects on environment, soil, and water. The overuse of biochar may disrupt soil structure by causing an imbalance in the soil's liquid and gas phases (Castellini et al. 2015). Soil fauna and carbon content may get lowered and put adverse impact on wheat growth (Ji et al. 2022). Therefore, it is essential for conducting more research of how biochar application will interact with crop yielding parameters and its related mechanisms.

10.5 Conclusion

Rice straw, a by-product of rice harvesting, poses significant environmental challenges due to its increasing production and the prevalent practice of open burning for disposal. The conversion of rice straw into biochar has come out as a potential solution to address the environmental issues related with its disposal. Biochar production from rice straw through pyrolysis has gained attention for its energy efficiency and environmental benefits. RSB offers several advantages as a soil amendment, including carbon neutrality, slow-release effects of nutrients, and improved soil water content and porosity. It has the ability to come up with the ecological and sustainable development of ecosystems. Additionally, RSB has been found effective in treatment of wastewater and the elimination of soil contaminants. This dual functionality of RSB as both a soil amendment and a sorbent expands its potential applications in various environmental remediation processes. The physical properties like moisture and bulk density are important factors in its handling and storage. Processing methods like pelletization and briquetting can increase its density and reduce its volume, facilitating easier storage and transportation. The thermal properties of rice straw, including calorific value and volatile matter, influence its conversion into biofuels. Understanding these properties is crucial for optimizing the production of biochar and other energy applications.

Chemically, it is composed of lignocellulose, with cellulose, hemicellulose, and lignin as major components. The quantitative and qualitative variations of these polymers in rice straw depend on factors such as variety, season, and geographical location. Understanding the chemical composition of rice straw is essential for its effective utilization as a feedstock for biochar production and as a soil fertility enhancer. While biochar production from rice straw shows promise, concerns regarding its greenhouse gas (GHG) mitigation potential need to be addressed. Further research is desired to ameliorate the process of biochar production and enhance its environmental benefits. Understanding the life cycle assessment of biochar production from rice straw will provide valuable insights into its overall environmental impact and its potential role in mitigating climate change. In conclusion, the utilization of rice straw as biochar offers a sustainable and efficient way to manage agricultural waste, mitigate environmental issues, and improve soil fertility, and the multifunctional nature of rice straw biochar as a soil amendment and sorbent expands its potential applications in various environmental remediation processes. However, further research and development are required to optimize the production process, address GHG mitigation concerns, and explore the full potential of rice straw biochar in different applications.

References

Abdelaziz OY, Brink DP, Prothmann J, Ravi K, Sun M, García-Hidalgo J et al (2016) Biological valorization of low molecular weight lignin. Biotechnol Adv 34(8):1318–1346

Abe K, Yano H (2009) Comparison of the characteristics of cellulose microfibril aggregates of wood, rice straw and potato tuber. Cellulose 16:1017–1023

Abitha A, Jesty CA, Siyad S, Vijay PC, Raj SS, Dhanya BS (2023) Feasibility assessment of rice straw ash as a supplementary cementitious material. Mater Today Proc. https://doi.org/10.1016/j.matpr.2023.01.165

Agarwal M (2014) An investigation on the pyrolysis of municipal solid waste. Thesis, RIMT

Ahamer G (2022) Why biomass fuels are principally not carbon neutral. Energies 15(24):9619

Ahmad M, Rajapaksha AU, Lim JE, Zhang M, Bolan N, Mohan D et al (2014) Biochar as a sorbent for contaminant management in soil and water: a review. Chemosphere 99:19–33

Al Arni S (2018) Comparison of slow and fast pyrolysis for converting biomass into fuel. Renew Energy 124:197–201

Bakraoui M, Karouach F, Ouhammou B, Aggour M, Essamri A, El Bari H (2020) Biogas production from recycled paper mill wastewater by UASB digester: optimal and mesophilic conditions. Biotechnol Rep 25:e00402

Barus J, RrErnawati RE, Wardani N, Pujiharti Y, Suretno ND, Slameto S (2023) Improvement in soil properties and soil water content due to the application of rice husk biochar and straw compost in tropical upland. Int J Recycl Org Waste Agric 12(1):85–95

Bezerra TL, Ragauskas AJ (2016) A review of sugarcane bagasse for second-generation bioethanol and biopower production. Biofuels Bioprod Biorefin 10(5):634–647

Binod P, Sindhu R, Singhania RR, Vikram S, Devi L, Nagalakshmi S et al (2010) Bioethanol production from rice straw: an overview. Bioresour Technol 101(13):4767–4774

Cao X, Ma L, Liang Y, Gao B, Harris W (2011) Simultaneous immobilization of lead and atrazine in contaminated soils using dairy-manure biochar. Environ Sci Technol 45(11):4884–4889

Castellini M, Giglio L, Niedda M, Palumbo AD, Ventrella D (2015) Impact of biochar addition on the physical and hydraulic properties of a clay soil. Soil Tillage Res 154:1–13

Cui C, Zhao S, Zhang Z, Li M, Shi R, Sun Q (2023) Preparation and characterization of corn starch straws with strong mechanical properties by extrusion and retrogradation. Ind Crop Prod 191:115991

Dai J, Meng X, Zhang Y, Huang Y (2020) Effects of modification and magnetization of rice straw derived biochar on adsorption of tetracycline from water. Bioresour Technol 311:123455

Das T, Thakur R, Dhua S, Teixeira-Costa BE, Rodrigues MB, Pereira MM et al (2023) Processing of cereals. In: Cereal grains. CRC Press, London, pp 195–223

Demirbas A (2004) Combustion characteristics of different biomass fuels. Prog Energy Combust Sci 30(2):219–230

Dobermann ATHF, Fairhurst TH (2002) Rice straw management. Better Crops Int 16(1):7–11

Enaime G, Baçaoui A, Yaacoubi A, Lübken M (2020) Biochar for wastewater treatment—conversion technologies and applications. Appl Sci 10(10):3492

Fang J, Zhan L, Ok YS, Gao B (2018) Minireview of potential applications of hydrochar derived from hydrothermal carbonization of biomass. J Ind Eng Chem 57:15–21

Gadde B, Bonnet S, Menke C, Garivait S (2009a) Air pollutant emissions from rice straw open field burning in India, Thailand and The Philippines. Environ Pollut 157(5):1554–1558

Gadde B, Menke C, Wassmann R (2009b) Rice straw as a renewable energy source in India, Thailand, and The Philippines: overall potential and limitations for energy contribution and greenhouse gas mitigation. Biomass Bioenergy 33(11):1532–1546

Goodman BA (2020) Utilization of waste straw and husks from rice production: a review. J Bioresou Bioprod 5(3):143–162

Hallac BB, Ragauskas AJ (2011) Analyzing cellulose degree of polymerization and its relevancy to cellulosic ethanol. Biofuels Bioprod Biorefin 5(2):215–225

Heidenreich S, Foscolo PU (2015) New concepts in biomass gasification. Prog Energy Combust Sci 46:72–95

Huang J, Liu C, Tong H, Li W, Wu D (2012) Theoretical studies on pyrolysis mechanism of xylopyranose. Comput Theor Chem 1001:44–50

Ibrahim M, Khan S, Hao X, Li G (2016) Biochar effects on metal bioaccumulation and arsenic speciation in alfalfa (Medicagosativa L.) grown in contaminated soil. Int J Environ Sci Technol 13:2467–2474

Jain A, Balasubramanian R, Srinivasan MP (2016) Hydrothermal conversion of biomass waste to activated carbon with high porosity: a review. Chem Eng J 283:789–805

Ji M, Wang X, Usman M, Liu F, Dan Y, Zhou L et al (2022) Effects of different feedstocks-based biochar on soil remediation: a review. Environ Pollut 294:118655

Kargbo FR, Xing J, Zhang Y (2010) Property analysis and pretreatment of rice straw for energy use in grain drying: a review. Agric Biol J N Am 1(3):195–200

Kausar H, Ismail MR, Saud HM, Berahim Z, Habib SH, Othman R, Bhuiyan SH (2016) Microbial composting of rice straw for improved stability and bioefficacy. In: Hakeem K, Akhtar M, Abdullah S (eds) Plant, soil and microbes, Implications in crop science, vol 1. Springer, Cham, pp 271–290

Khorram MS, Zhang Q, Lin D, Zheng Y, Fang H, Yu Y (2016) Biochar: a review of its impact on pesticide behavior in soil environments and its potential applications. J Environ Sci 44:269–279

Klimiuk E, Pawłowski M, Pokój T (2012) Biofuels technologies for sustainable development. PWN SA, Warsaw

Koppejan J, Sokhansanj S, Malin S, Madrali S (2012) Status overview of torrefaction technologies. IEA Bioenergy Task 32 Raport, Enschede

Kumar S, Kalra S, Sahni T, Sidhu SK, Sarao LK (2023) Agricultural lignocellulosic waste to biofuels. In: Srivastava N, Verma B, Mishra PK (eds) Agroindustrial waste for green fuel application. Springer Nature, Singapore, pp 205–247

Lee J, Lee K, Sohn D, Kim YM, Park KY (2018) Hydrothermal carbonization of lipid extracted algae for hydrochar production and feasibility of using hydrochar as a solid fuel. Energy 153: 913–920

Li J, Hu Z, Li F, Fan J, Zhang J, Li F, Hu H (2019) Effect of oxygen supply strategy on nitrogen removal of biochar-based vertical subsurface flow constructed wetland: intermittent aeration and tidal flow. Chemosphere 223:366–374

Li H, Liu Y, Jiao X, Li J, Liu K, Wu T et al (2023) Response of soil nutrients retention and rice growth to biochar in straw returning paddy fields. Chemosphere 312:137244

Lin Y, Ma X, Peng X, Yu Z, Fang S, Lin Y, Fan Y (2016) Combustion, pyrolysis and char CO_2-gasification characteristics of hydrothermal carbonization solid fuel from municipal solid wastes. Fuel 181:905–915

Liu Z, Xu A, Long B (2011) Energy from combustion of rice straw: status and challenges to China. Energy Power Eng 3(03):325

Liu WJ, Jiang H, Yu HQ (2015) Development of biochar-based functional materials: toward a sustainable platform carbon material. Chem Rev 115(22):12251–12285

Lv R, Wang Y, Wang Q, Wen Y, Shang Q (2023) Rice straw biochar alters inorganic nitrogen availability in paddy soil mainly through abiotic processes. J Soils Sediments 23(2):568–581

Maiorella BE (1985) In: Moo-Young M (ed) Comprehensive biotechnology, vol 4. Pergamon, New York

Medha I, Chandra S, Bhattacharya J, Samal B, Vanapalli KR (2023) Development of rice straw-derived biochar-bentonite composite and its application for in situ sequestration of ammonium and phosphate ions in the degraded mine soil. Environ Manage 71(5):1065–1086

Mei Y, Li B, Fan S (2020) Biochar from rice straw for Cu^{2+} removal from aqueous solutions: mechanism and contribution made by acid-soluble minerals. Water Air Soil Pollut 231:1–13

Mendoza TC, Samson R (1999) Strategies to avoid crop residue burning in the Philippine context. Resource Efficient Agricultural Production (REAP), Canada

Migo MVP (2019) Optimization and life cycle assessment of the direct combustion of rice straw using a small scale, stationary grate furnace for heat generation. Unpublished masters thesis, University of the Philippines Los Baños

Ministry of Agriculture and Farmers Welfare (2021) Anuual report 2021. Departemnt of Agriculture, Cooperation and Farmers Welfare, Ministry of Agriculture and Farmers Welfare, Government of India

Morsy MI, Alakeel KA, Ahmed AE, Abbas AM, Omara AI, Abdelsalam NR, Emaish HH (2022) Recycling rice straw ash to produce low thermal conductivity and moisture-resistant geopolymer adobe bricks. Saudi J Biol Sci 29(5):3759–3771

Mu W, Ben H, Ragauskas A, Deng Y (2013) Lignin pyrolysis components and upgrading—technology review. Bioenergy Res 6:1183–1204

Ngamchompoo W (2018) Experimental investigation on rice straw gasification in a cyclone gasifier. BioResources 13(4):7654–7670

Nghi NT, Romasanta RR, Van Hieu N, Vinh LQ, Du NX, Ngan NVC et al (2020) Rice straw-based composting. In: Gummert M, Hung N, Chivenge P, Douthwaite B (eds) Sustainable rice straw management. Springer, Cham, pp 33–41

Nhuchhen DR, Basu P, Acharya B (2014) A comprehensive review on biomass torrefaction. Int J Renew Energy Biofuels 2014:1–56

Nkosi N, Muzenda E, Gorimbo J, Belaid M (2021) Developments in waste tyre thermochemical conversion processes: gasification, pyrolysis and liquefaction. RSC Adv 11(20):11844–11871

Osayi JI, Iyuke S, Ogbeide SE (2014) Biocrude production through pyrolysis of used tyres. J Catal 2014:386371

Palaniappan SP (2006) Cropping Systems in the Tropics (principles and management). New Age International, New Delhi

Pandey SM, Kudesia S, Pandey J, Singh U (2023) A sustainable approach for soil amendment; available plant nutrients in rice straws. J Indian Chem Soc 100:100938

Pang S (2019) Advances in thermochemical conversion of woody biomass to energy, fuels and chemicals. Biotechnol Adv 37(4):589–597

Prabakar D, Manimudi VT, Sampath S, Mahapatra DM, Rajendran K, Pugazhendhi A (2018) Advanced biohydrogen production using pretreated industrial waste: outlook and prospects. Renew Sust Energ Rev 96:306–324

Qu K, Huang L, Hu S, Liu C, Yang Q, Liu L et al (2023) TiO_2 supported on rice straw biochar as an adsorptive and photocatalytic composite for the efficient removal of ciprofloxacin in aqueous matrices. J Environ Chem Eng 11:109430

Rani J, Paul B (2023) Challenges in arid region reclamation with special reference to Indian Thar Desert—its conservation and remediation techniques. Int J Environ Sci Technol 20:12753–12774. https://doi.org/10.1007/s13762-022-04746-z

Rathour RK, Devi M, Dahiya P, Sharma N, Kaushik N, Kumari D et al (2023) Recent trends, opportunities and challenges in sustainable management of rice straw waste biomass for green biorefinery. Energies 16(3):1429

Roberto IC, Mussatto SI, Rodrigues RC (2003) Dilute-acid hydrolysis for optimization of xylose recovery from rice straw in a semi-pilot reactor. Ind Crop Prod 17(3):171–176

Sakhiya AK, Vijay VK, Kaushal P (2023) Development of rice straw biochar through pyrolysis to improve drinking water quality in arsenic and manganese contaminated areas. Surf Interfaces 36:102582

Seow YX, Tan YH, Mubarak NM, Kansedo J, Khalid M, Ibrahim ML, Ghasemi M (2022) A review on biochar production from different biomass wastes by recent carbonization technologies and its sustainable applications. J Environ Chem Eng 10(1):107017

Shanmugam V, Sreenivasan SN, Mensah RA, Försth M, Sas G, Hedenqvist MS et al (2022) A review on combustion and mechanical behaviour of pyrolysis biochar. Mater Today Commun 31:103629

Shuai L, Amiri MT, Questell-Santiago YM, Héroguel F, Li Y, Kim H et al (2016) Formaldehyde stabilization facilitates lignin monomer production during biomass depolymerization. Science 354(6310):329–333

Stępień P, Pulka J, Białowiec A (2017) Organic waste torrefaction–a review: reactor systems, and the biochar properties. In: Samer M (ed) Pyrolysis. IntechOpen, London, pp 37–52

Takkellapati S, Li T, Gonzalez MA (2018) An overview of biorefinery-derived platform chemicals from a cellulose and hemicellulose biorefinery. Clean Techn Environ Policy 20:1615–1630

Van Hung N, Maguyon-Detras MC, Migo MV, Quilloy R, Balingbing C, Chivenge P, Gummert M (2020) Rice straw overview: availability, properties, and management practices. In: Gummert M, Hung N, Chivenge P, Douthwaite B (eds) Sustainable rice straw management. Springer, Cham, pp 1–13

Vu T, Nguyen HS (2011) Application of biochar to improve soil productivity—affecting the type and amount of biochar applied to rice growth and productivity. Vietnam J Agric Sci Technol 1: 56–60

Wang Y, Yin R, Liu R (2014) Characterization of biochar from fast pyrolysis and its effect on chemical properties of the tea garden soil. J Anal Appl Pyrolysis 110:375–381

Wang X, Zhong Z, Jin B (2020) Experimental evaluation of biomass medium-temperature gasification with rice straw as the fuel in a bubbling fluidized bed Gasifier. Int J Chem React Eng 18(1):20190147. https://doi.org/10.1515/ijcre-2019-0147

Wang S, Guo X, Lu H, Liu H (2023) Effect of powder column height on the compression behavior of rice husk powder under gas pressurization. Fuel 340:127615

Warnock DD, Lehmann J, Kuyper TW, Rillig MC (2007) Mycorrhizal responses to biochar in soil–concepts and mechanisms. Plant Soil 300:9–20

Wei J, Tu C, Yuan G, Liu Y, Bi D, Xiao L et al (2019) Assessing the effect of pyrolysis temperature on the molecular properties and copper sorption capacity of a halophyte biochar. Environ Pollut 251:56–65

Xia L, Chen W, Lu B, Wang S, Xiao L, Liu B et al (2023) Climate mitigation potential of sustainable biochar production in China. Renew Sust Energ Rev 175:113145

Xu X, He C, Yuan X, Zhang Q, Wang S, Wang B et al (2020) Rice straw biochar mitigated more N_2O emissions from fertilized paddy soil with higher water content than that derived from ex situ biowaste. Environ Pollut 263:114477

Yang B, Dai Z, Ding SY, Wyman CE (2011) Enzymatic hydrolysis of cellulosic biomass. Biofuels 2(4):421–449

Young GC (2010) Municipal solid waste to energy conversion processes: economic, technical, and renewable comparisons. Wiley, Hoboken

Zeleke TS, Yihun FA, Ayana MT, Kassa MT, Alemante MF (2023) Enhancing the thermo-mechanical properties of thermoplastic starch films using rice straw fibers as reinforcement. Chem Afr 6:2321–2329

Zhang Q, Li Q, Zhang L, Yu Z, Jing X, Wang Z et al (2017) Experimental study on co-pyrolysis and gasification of biomass with deoiled asphalt. Energy 134:301–310

SPRINGER NATURE

GPSR Compliance

The European Union's (EU) General Product Safety Regulation (GPSR) is a set of rules that requires consumer products to be safe and our obligations to ensure this.

If you have any concerns about our products, you can contact us on ProductSafety@springernature.com

In case Publisher is established outside the EU, the EU authorized representative is:

Springer Nature Customer Service Center GmbH
Europaplatz 3
69115 Heidelberg, Germany

The manufacturer's authorised representative in the EU is Springer Nature Customer Service Centre GmbH, Europaplatz 3, 69115 Heidelberg, Germany. If you have any concerns regarding our products, please contact ProductSafety@springernature.com

Printed and bound by CPI Group (UK) Ltd, Croydon, CR0 4YY
30/09/2026
02225322-0003